PLANT GENOMICS

Edited by:

Hany A. El-Shemy
University of Cairo
Egypt

This edition published by Caister Academic Press 2018

First published in 2009 by Savanna Press

British Library Cataloguing-in-Publication Data

A catalogue record for this book is available from the British Library

ISBN: 978-1-912530-06-9

Description or mention of instrumentation, software, or other products in this book does not imply endorsement by the author or publisher. The author and publisher do not assume responsibility for the validity of any products or procedures mentioned or described in this book or for the consequences of their use.

Contents

Contributors

- Amr F. Abdelkhalik, Rice Research and Training Center, Egypt
- Ahmed M. Aboul-Enein, University of Cairo, Egypt
- Mourad A. M. Aboul-Soud, University of Cairo, Egypt
- A.J. Afzal, Southern Illinois University, Carbondale, USA
- R. Ahsan, State University, Blacksburg, USA
- Liakat Ali, University of Guelph, Canada
- Megahed H. Ammar, Rice Research and Training Center, Egypt
- Kotb A. Attia, Rice Research and Training Center, Egypt
- Shimaa Bahgat, University of Cairo, Egypt
- Sukesh Bhaumik, Southern Illinois University, Carbondale, USA
- Wagih M. El-Sayed, Alexandria University, Egypt
- Hany A. El-Shemy, University of Cairo, Egypt
- Osama El-Shihy, University of Cairo, Egypt
- Deepak Ganjewala, VIT University, India
- L. Gupta, Southern Illinois University, Carbondale, USA
- Muhammad Haroon, Potato Research Centre, Canada
- M.J. Iqbal, State University, Blacksburg, USA
- Masao IshimotoNational Agricultural Research Center, Japan
- Michiaki Iwata, Meiji Seika Kaisha, Ltd, Japan
- A. Jamai, Southern Illinois University, Carbondale, USA
- U.B. Karangula, Southern Illinois University, Carbondale, USA
- M.A. Kassem, Fayetteville State University, Fayetteville, USA
- Mutasim M. Khalafalla, National Centre for Research, Sudan
- Shiv Kumar, VIT University, India
- Xiu-Qing Li, Potato Research Centre, Canada
- David A. Lightfoot, Southern Illinois University, Carbondale, USA
- Jie Liu, University of Guelph, Canada
- Qiang Liu, Guelph Food Research Centre, Ontario, Canada
- Rajesh Luthra, Council of Scientific and Industrial Research, India
- Aparna Natarajan, Southern Illinois University, Carbondale, USA
- Eman A. Mahmoud, National Research Centre, Egypt
- K. Meksem, Southern Illinois University, Carbondale, USA
- Solliman M. Mohei El-Din, National Research Centre, Egypt
- Yvan Pelletier, Potato Research Centre, Canada
- K. Peter Pauls, University of Guelph, Canada
- Omer A. Shabban, University of Cairo, Egypt
- Abhijit Shukla, Southern Illinois University, Carbondale, USA
- Ghanem A. SobhyNational Research Centre, Egypt
- Jeff Taylor, University of Guelph, Canada
- Jennifer Tedman, University of Guelph, Canada
- Chun Wei, Fudan University, China
- Jinshui Yang, Fudan University, China
- Jiazheng Yuan, University of Guelph, Canada

Chapter 1

A High-Resolution Melting Approach for Analyzing Allelic Expression Dynamics

Jiazheng Yuan, Muhammad Haroon, David A. Lightfoot, Yvan Pelletier, Qiang Liu, and Xiu-Qing Li

Abstract

Single nucleotide polymorphisms (SNPs) are single base pair mutations that provide new approaches to studies of allele transcript abundances. High-resolution DNA melting curve (HRM) analysis using a LightScanner (Hi-Res Melting™ system with Idaho's LC Green) provides post-PCR detection of mutations and SNPs in genomic DNA. This study investigated whether the HRM analysis can distinguish alleles among potato (*Solanum tuberosum*) transcript abundances. Transcript properties of genes encoding seven carbohydrate metabolism enzymes/proteins in various tissues and cold storage durations were studied. The HRM assay measured differential expression of alleles between different organs, between different storage treatments and stages of tubers from the same variety, and between different varieties with the same treatment. The RT-PCR amplicons were directly sequenced to assist the interpretation of HRM data. The cDNA HRM curves correlated well with the nucleotide polymorphisms of the cDNA sequences and the transcript abundance of alleles and therefore can serve as functional allele activity (FAA) markers. By combining the allelic specificity of HRM with simple PCR design, this technology can be applied to rapidly determine the most active allele of a gene among the cells analyzed.

Introduction

Trait differences among cultivars are mainly due to the inheritance and or expression of different alleles (Foolad and Lin, 1998). Because most genes have multiple alleles, many produce multiple nascent transcripts to respond coordinately to the temporal, spatial, or external stimuli.

The melting temperature of a double strand DNA fragment is defined as the temperature where 50% of the DNA is single stranded (SantaLucia Jr, 1998). The nucleotide CG-pairs have a higher melting temperature tolerance than AT-pairs, and the latter has a higher tolerance than deletions. If an allele has a higher proportion of CG pairs than another allele, the allele shows a higher melting temperature. Melting temperature is one of the characteristics of DNA, and cDNA and can be used in describing the difference (polymorphism) among alleles (Howell *et al.*, 2002; Wittwer *et al.*, 2003). Single nucleotide polymorphisms (SNPs) are single base pair mutations at specific loci. They are the most common type of genetic variation and also evolutionarily conserved (Syvanen, 2001). Therefore, SNP-based markers can be used to facilitate quantitative trait loci (QTL) analysis and trait association studies. High-resolution DNA melting curve analysis has become a robust method to perform genotyping (Montgomery *et al.*, 2007) and provides sensitive detection of mutations, including SNPs.

Gene transcriptional regulation has been studied in several ways, including Northern hybridization (Alwine *et al.*, 1977), real time PCR (Nolan *et al.*, 2006), serial analysis of gene expression (SAGE) (Velculescu *et al.*, 1995), expressed sequence tag (EST) sequencing (Flinn *et al.*, 2005), and microarray analysis (Meyers *et al.*, 2004). Each of these technologies provides a measurement of gene expression with various strengths and shortcomings. Transcriptional changes play a major role in many plant physiological processes. Analysis of transcription abundances allows the inference of coordinated transcription events and organized gene expressions. Dissection of alterations in transcription abundance has provided unique opportunities to study gene function by the comparison of species, tissues, and time durations simultaneously (Schenk *et al.*, 2000; Zhao *et al.*, 2004; Iqbal *et al.*, 2005; Journot-Catalino *et al.*, 2006). To detect and distinguish several transcripts sensitively, methods such as fluorescence dye and distinct color labeled oligo DNA probes have been used to detect many transcripts using the real-time PCR approach (Jeong *et al.*, 2007). These spectral approaches usually possess a minimum of two distinct fluorophores to increase specificity. However, the assays are laborious and often do not contain specific or allelic information during the transcript analysis.

A simple approach that can sensitively visualize expressed allelic variants and the transcript abundance dynamics in high throughput using a small amount of RNA is still a challenge. Few current techniques for gene expression assays allow the analysis of transcript abundance regulation in a structural context (Guo *et al.*, 2006). Real-time PCR analysis is another approach, but its resolution often does not reach a single nucleotide level during transcript analysis. The LightScanner is a post-PCR Hi-Res Melting™ system (Idaho Technology

Inc, Salt Lake City, UT) for genotyping and mutation scanning. The system has been widely used in genomic DNA genotyping. The LightScanner produces better scanning specificity than current popular real-time Q-PCR instruments (Herrmann *et al.*, 2007). The dsDNA binding dyes LCGreen® Plus offers better sensitivity and specificity in the DNA melting analysis than that of SYBR Green (Wittwer *et al.*, 2003; Herrmann *et al.*, 2006; Herrmann *et al.*, 2007). The HRM approach enabled the detection of RNA editing of chloroplast transcripts by mixing plasmid DNA containing a C and plasmid DNA containing a T at the same site in various ratios (Chateigner-Boutin and Small, 2007). Each plant cell has many chloroplasts, chloroplast RNA is expected to be in a higher density than the mRNA density of a nuclear gene. Chloroplast RNA can be quite extensively edited along the transcript molecules and can have a 3.6% change amino acid sequence identity (Kugita *et al.*, 2003). Therefore, chloroplast RNA HRM assay does not necessarily conclude whether HRM can effectively monitor nuclear gene allele activities and transcript properties.

Here, the LightScanner-based HRM was employed to study potato nuclear gene derived transcript abundance; to infer coordinated transcription events; and to study the allelic dynamics of gene expression within various organs, tissues, genotypes, and treatment. The allelic variants of genes accumulated in temporal and spatially distinct fashions. By combining advances of fluorescence detection with simple PCR design, visualization of allelic expression dynamics was achieved. This technology has potential to monitor allele activation and silencing and to determine the changes of the dominant transcript in genetically heterogeneous cells. Thus, a new platform for the study of the potato transcriptome is demonstrated.

Methods

Plant materials and primer design

Potato varieties 'Shepody', 'Atlantic', and 'Superior' were used. For the tissue specific expression of genes, 15 types of tissues of Shepody were studied, which included tuber tissues; perimedulla, vascular ring, cortex, and skin; and different organs of the plants; flowers, fruits, shoot tips, health fully opened leaves, old yellowing leaves, petioles, young stems, young roots, stolons, tiny young tubers, and mature tubers, young leaves, old leaves. For the study on tissue specificity, tissue/organ samples were from Shepody plants growing in the greenhouses. For storage stage gene expression analysis, Atlantic and Superior plants were grown in the field, and the mature tubers were stored at the storage facility (7°C) at the Potato Research Centre of Agriculture and Agri-Food Canada (Fredericton, NB).

After the preliminary screening, five genes (seven primer pairs) were studied (Table 1), and three genes including acid invertase, acid invertase inhibitor, and ADP-glucose pyrophosphorylase were studies in detail. All oligonucleotide primers were designed based on EST sequences surrounding SNP regions of a known genotype (Shepody). The negative control was the non-reverse transcribed RNA. The amplicons ranged in size between 100–300 bp, and the Tm of the PCR products were estimated to be less than 95°C to ensure the observation of the entire scope of melting curves.

Nucleic acid isolation

Plant tissues were immediately frozen in liquid nitrogen and kept at –80°C. Total RNA was isolated using the RNeasy plant mini kit (Qiagen, Mississauga, ON) in combination with DNase treatment using the RNase-free DNase kit (Qiagen GmbH, Hilden, Germany) according to the manufacturer's instruction.

PCR conditions and HRM analysis

Reverse transcription was carried out with 1 µg of total RNA with the Retroscript kit at 50°C for one hour and 92°C for 5 min, following manufacturer's instruction to prepare cDNA (Stratagene, La Jolla, CA). All of amplicons were amplified from potato cDNA in 10 µl PCR reactions containing 20 ng cDNA template, PCR buffer (1×), 1 U of Taq polymerase (Invitrogen, Burlington, ON), 2 mM $MgCl_2$, 1 mM each of dATP, dCTP, dTTP and dGTP, 1 µl of LC Green Plus (Idaho Technology, Salt Lake City, UT) and 20

Table 1. List of primers used to screen various tissues for the tissue- and treatment-specific cDNAs.

Primer name	Gene	Sequence
CC1533A	ADP-glucose pyrophosphorylase	CC1533A-L: ACAAAGACAACGTTCAAGAAG CC1533A-R: CAGCACGTTTCTACCTCATG
CA548A	Acid invertase	CA548A-L: AGGACAGTGCTTTACGACAA CA548A-R: CTCTGCAGCTGAGTCAACA
CA548B	Acid invertase	CA548B-L: ATTGAAGCAGATCATGTAGG CA548B-R: ATCAGCACAGAAGTGAGTCT
CP7079A	Glutamic acid decarboxylase	CP7079A-L: CTAACGTCTCAGTTGTGGAC CP7079A-R: CTTAACCATGTCATCTCCCAT
CA35A	Invertase inhibitor	CA35A-L: CACATGTCGATTCCCCAAT CA35A-R: GTCGTTTTCATTTCCTACGT
CA35B	Invertase inhibitor	CA35B-L: TGGGAAAAGGCACGCC CA35B-R: AGATGAGAGGGACAATGCAG
CP96	Sucrose synthase	CP96-L: CACTTGGCAAATCTACTCCG CP96-R: CATCTTACGGTACTTGAGAGC

pmol of the primers. Primers were synthesized according to the sequences designed for SNPs in Shepody ESTs (Griffiths *et al.* http://10.110.16.26/SNP/snp.html). PCR amplifications were performed in a DNA Thermo-cycler (MJ Research Inc., Scarborough, ON) programmed for 41 cycles of 1 min at 95°C, 1 min at 50–65°C and 1 min at 72°C. The PCR experiments were repeated at least once. Following amplification, the melting acquisition was performed on a LightScanner instrument (Idaho technology). Melting curve analysis was carried out with LightScanner software (version 2.0). Selected amplicons were purified with QIAquick PCR purification kit (Qiagen, Mississauga, ON) and directly sequenced by using gene specific primers with ABI Prism 3100 Genetic Analyzer facilitated in Institut de recherches cliniques de Montreal (IRCM; Montreal, QC) and Research Productivity Council (Fredericton, NB). Sequence graphs were visualized using the BioEdit program (Hall, 2004, http://www.mbio.ncsu.edu/BioEdit/bioedit.html). ClustalX2 was used to align the sequences and to generate Bootstrap-1000 NJ phylogeny trees. The LightScanner data were normalized by both local (local background value was subtracted from the intensity value of sample) and global metrics.

Results

Correlation between HRM curve characteristics and transcript properties in different potato tissues

All the 7 pairs of primers listed in Table 1 worked for the reverse transcription PCR (RT-PCR) and for the generation of HRM curves in the test stage of the primer using a few samples (data not shown). Full experiments were conducted only using 75 tissue-primer combinations, which resulted from PCR amplification with only five pairs of primers and 15 organs/tissues of the potato variety Shepody. The 5 primer pairs were derived from the cDNA sequences of acid invertase (EC 3.2.1.26), invertase inhibitor (two pairs), ADP glucose pyrophosphorylase (AGPase; EC 2.7. 7.27), and sucrose synthase (EC 2.4.1.13). HRM analysis of these four genes in the various potato tissues from the variety Shepody revealed unique patterns of HRM curves. All four of these assayed genes contained more than one sequence variants in these different organs and tissues. Figs 1 and 4 showed the cDNA analyzed with two pairs of the primers designed for the acid invertase (CA548B) and the invertase inhibitor (CA35B) genes.

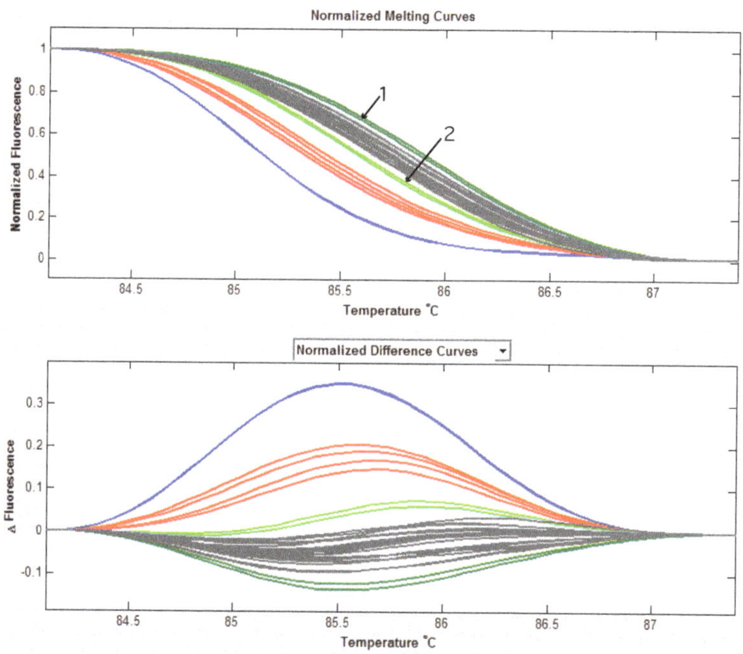

Fig. 1. High resolution melting curves of allelic expression variants of the invertase inhibitor gene locus (primer pair CA35B). Variety; Shepody; dark green curves: Young roots; light green curve: Young stems and old leaves; grey curves: Fruit, petiole, stolons, young tuber, mature tuber, skin of tuber, cortex, vascular ring and perimedullar; red curves: Shoot tip and young leaves; blue curves: Open flower; not shown. The PCR reactions for each tissue were performed in duplicate. Melting curve normalization was carried out with LightScanner software on the LightScanner (Idaho Inc., UT, USA).

```
         1        10        20        30        40        50
TuberY1  TGGTGATGTGGTGCGTGTAATGACTGGTGTTTATATCAAGTTCTTTGTTA
TuberY2  TGGTGATGTGGTGCGTGTAATGACTGGTGTTTATATCAAGTTCTTTGTTA
TuberM1  TGGTGATGTGGTGCGTGTAATGACTGGTGTTTATATCAAGTTCTTTGTTA
TuberM2  TGGTGATGTGGTGCGTGTAATGACTGGTGTTTATATCAAGTTCTTTGTTA
Fruit1   TGGTGATGTGGTGCGTGTAATGACTGGTGTTTATATCAAGTTCTTTGTTA
Fruit2   TGGTGATGTGGTGCGTGTAATGACTGGTGTTTATATCAAGTTCTTTGTTA
Root1    TGGTGATGTGGTGCGTGTAATGACTGTTGTTTATATCAAGTTCTTTGTTA
Root2    TGGTGATGTGGTGCGTGTAATGACTGTTGTTTATATCAAGTTCTTTGTTA
         ************************** ***********************

TuberY1  AAACATCAAAGTTGTGTGTTGACCAAACTGTTTGGAAAGTTAATCATGAA
TuberY2  AAACATCAAAGTTGTGTGTTGACCAAACTGTTTGGAAAGTTAATCATGAA
TuberM1  AAACATCAAAGTTGTGTGTTGACCAAACTGTTTGGAAAGTTAATGATGAA
TuberM2  AAACATCAAAGTTGTGTGTTGACCAAACTGTTTGGAAAGTTAATGATGAA
Fruit1   AAACAACAAAGTTGTGTGTTGACCAAACTGTTTGGAAAGTTAATAATGAA
Fruit2   AAACAACAAAGTTGTGTGTTGACCAAACTGTTTGGAAAGTTAATAATGAA
Root1    AAACAACAAAGTTGTGTGTTGACCAAACTGTTTGGAAAGTTAATGATGAA
Root2    AAACAACAAAGTTGTGTGTTGACCAAACTGTTTGGAAAGTTAATGATGAA
         ***** ************************************** *****

TuberY1  CAGTTGGTGGTAACTGGTGGTAACGTAGGAAATGAAAACGACATCTTCAA
TuberY2  CAGTTGGTGGTAACTGGTGGTAACGTAGGAAATGAAAACGACATCTTCAA
TuberM1  CAGTTGGTGGTAACTGGTGGTAACGTAGGAAATGAAAACGACATCTTCAA
TuberM2  CAGTTGGTGGTAACTGGTGGTAACGTAGGAAATGAAAACGACATCTTCAA
Fruit1   CAGTTGGTGGTAACTGGTGGTAACGTAGGAAATGAAAACGACATCTTCAA
Fruit2   CAGTTGGTGGTAACTGGTGGTAACGTAGGAAATGAAAACGACATCTTCAA
Root1    CAGTTGGTGGTAACTGGTGGTAAGGTAGGAAATGAAAACGACATCTTCAA
Root2    CAGTTGGTGGTAACTGGTGGTAAGGTAGGAAATGAAAACGACATCTTCAA
         *********************** **************************

TuberY1  GATCAAGAAAACTGATTTGGTGATACCAGG---TATGAAAAATGTATACA
TuberY2  GATCAAGAAAACTGATTTGGTGATACCAGG---TATGAAAAATGTATACA
TuberM1  GATCAAGAAAACTGATTTGGTGATACCAGG---TATGAAAAATGTATACA
TuberM2  GATCAAGAAAACTGATTTGGTGATACCAGG---TATGAAAAATGTATACA
Fruit1   GATCAAGAAAACTGACTTGGTGATACCAGG---TATGAAAAATGTATACA
Fruit2   GATCAAGAAAACTGACTTGGTGATACCAGG---TATGAAAAATGTATACA
Root1    GATTATGAAAACTGACTTGGTGACACCAGGAGGTTCCAAATATGTATACA
Root2    GATTATGAAAACTGACTTGGTGACACCAGGAGGTTCCAAATATGTATACA
         *** * ********* ******* ******   *   *** *********
```

Fig. 2. 'Shepody' cDNA sequence alignments of the most abundantly expressed variants (alleles). The primer (CA35B) used to screen each tissue was derived from invertase inhibitor. The purified amplicons were sequenced by an ABI Prism 3100 Genetic Analyzer. *: Consensus sequences among the allele transcripts.

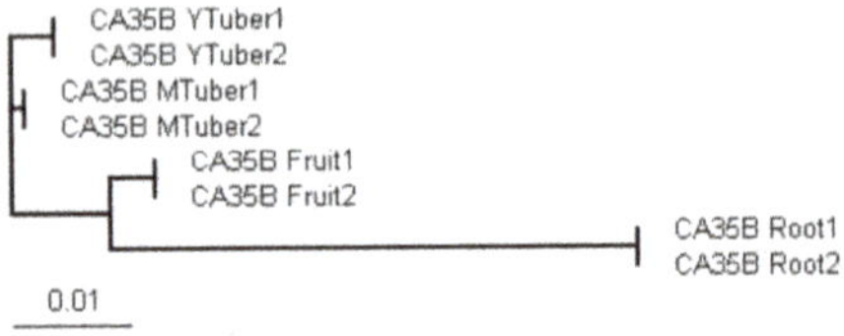

Fig. 3. Sequence-based phylogeny tree of the major transcripts detected by RT-PCR using the invertase inhibitor primer pair CA35B and RNA of various tissues. The scale bar represents an evolutionary distance of 0.01 nucleotide substitution per position.

The distinction of the melting curves was also confirmed by direct sequencing of RT-PCR amplicons. The sequences of some amplicons in several organs and tissues (e.g., Fig. 1: young-stem-CA35B) could not be defined because of the interference of non-ignorable basal expression from other polymorphic alleles or paralogs; However, there was a dominant variant in each sample at most times. It is known that the melting temperatures of different base pairs follow the order of CG>AT deletion and that a mixture of transcripts of the two would bring the melting temperature of the PCR products towards the mean value. Therefore, in most cases, HRM cures can suggest whether it is the higher Tm allele or the low Tm allele being the most actively expressed one in each RNA sample and can monitor whether the most active allele is a different one at different developmental stages, without the actual sequencing of the PCR products, as long as the genomic sequence of the SNP region is known.

HRM cures can reflect the characteristics of most abundant transcripts in multiple sample comparisons and can detect the allelic variation dynamics. Several PCR products from the heterozygous tetraploid potato variety Shepody were sequenced (Fig. 2) and a phylogenetic tree (Fig. 3) was generated. The HRM curves (Fig. 1) gave very similar distribution, reflecting well the allele positions on the phylogeneic tree (Fig. 3). There are four alleles detected for this invertase inhibitor gene locus (Fig. 2). The amplicons from root cDNAs had a small insert of three bases and therefore were expected to have a highest Tm value than the cDNA amplicons from the other tissues and organs. Transcripts in young tubers, mature tubers, and fruits were different from each other but more or less similar (Fig. 2), compared to their differences with the root expressed variant allele. The phylogenetic tree showed that the root cDNA variant was considerably different from the cDNA sequences from the tuber and fruit tissues. The HRM curves perfectly reflected this difference in the phylogenetic tree and showed a particularly separation of the HRM curves of roots (dark green curves) from the curves for these tubers and fruits (grey curves in Fig. 1). In addition to this positional separation between the HRM curves and phylogenic tree, the HRM curves also indicated the difference of Tm value between the amplicons (Fig. 1 top panel). Therefore, the simple HRM assay can reflect well the sequence characteristics of the transcripts expressed from the alleles at the same locus.

Sensitive reflection of HRM curves to the activity ratio between alleles

In the analysis of the acid invertase (CA548B), the HRM curves showed a clear difference between the young roots and other tissues (stolons, mature tubers, and vascular ring tissues) (Fig. 4). The cDNA curves of young roots showed a higher Tm value by about 0.5°C (Fig. 4). However, the DNA sequence reading from direct sequencing of the RT-PCR products did not show the expected polymorphism at the T/C SNP (the bold T on the right side of Fig. 5). There is one base difference in one of the sequencing repeats of vascular cDNA (the C/A SNP on the left side of Fig. 5), but did not have the same reading in the other repeat sequencing and did not solve the question why only the root HRM showed a higher Tm value. To solve this puzzle, we visually compared the raw sequence graphs between cDNA sequences from different tissues (Figs 6 and 7). It was found that the C-allele was expressed only at the nearly background level (Fig. 6) for the T/C SNP marked at right side in Fig. 5. However, there was a C/A SNP (the marked SNP at the left side in Fig. 5) at approximately the sequence position 61 in the PCR product sequences (Fig. 7). Both C-allele and A-allele were expressed in all the tissues but the activity ratio is different. In the young roots, most transcripts are from the C-allele, and very little from the A-allele (Fig. 7 left panel); whereas, in other tissues the

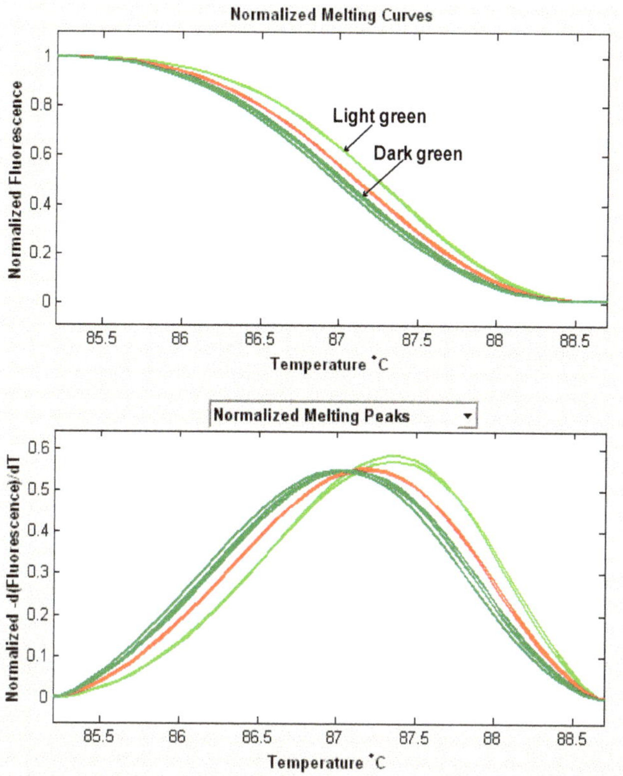

Fig. 4. High resolution melting curves of a SNP region (CA548B) of an acid invertase locus. Light green curve (marked by the arrow): Young roots; red curve: Stolons; dark green curves: Mature tuber and vascular ring. The PCR reactions for a tissue were amplified in duplicate. Melting curve normalization was carried out with LightScanner software on the LightScanner (Idaho Inc., UT). Note that the HRM curves of the two repeats of each tissue are very similar and that young root curves are different from the rest.

```
Y.Root1   CAAACGCTATCTGAGCTAACGCCAGTTTACTTCTACATTTCTAAAGGAGCTGATGGTCGAGC
Y.Root2   CAAACGCTATCTGAGCTAACGCCAGTTTACTTCTACATTTCTAAAGGAGCTGATGGTCGAGC
Stolon2   CAAACGCTATCTGAGCTAACGCCAGTTTACTTCTACATTTCTAAAGGAGCTGATGGTCGAGC
M.Tuber1  CAAACGCTATCTGAGCTAACGCCAGTTTACTTCTACATTTCTAAAGGAGCTGATGGTCGAGC
M.Tuber2  CAAACGCTATCTGAGCTAACGCCAGTTTACTTCTACATTTCTAAAGGAGCTGATGGTCGAGC
Vascu1    CAAACGCTATCTGAGCTAACGCCAGTTTACTTCTACATTTCTAAAGGAGCTGATGGTCGAGC
Vascu2    CAAAAGCTATCTGAGCTAACGCCAGTTTACTTCTACATTTCTAAAGGAGCTGATGGTCGAGC
          **** *********************************************************
```

Fig. 5. Sequence alignments (cDNA) showing the C/A SNP and T/C SNP region of the most abundantly expressed variants (alleles) of the acid invertase locus (primer CA548B) sequenced directly from RT-PCR products. Potato variety: Shepody. *: Consensus sequences among the allele transcripts. Note that the cDNA sequences readings of vascular tissue RT-PCR products are different; one is C, and the other reading is A.

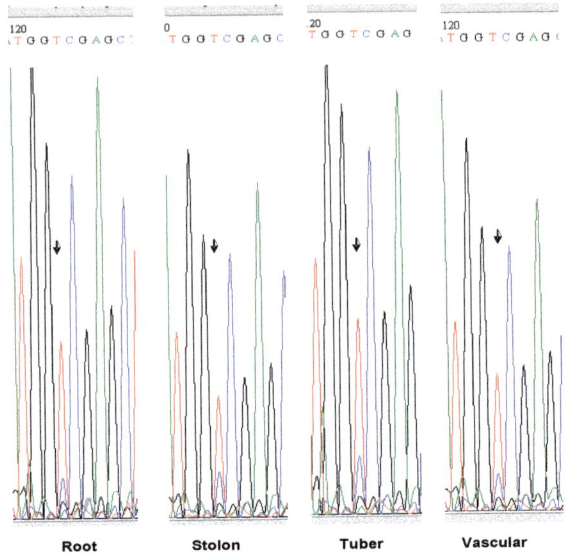

Fig. 6. Part of the raw cDNA sequence files of the targeted T/C SNP region of the abundantly expressed variants (alleles) at the acid invertase locus (primer CA548B) sequenced directly from RT-PCR products. Potato genotype: Shepody. Note that the C allele (blue peak at the SNP) transcript was at low abundance in all the four types of tissues and therefore is unlikely to influence the HRM curves significantly.

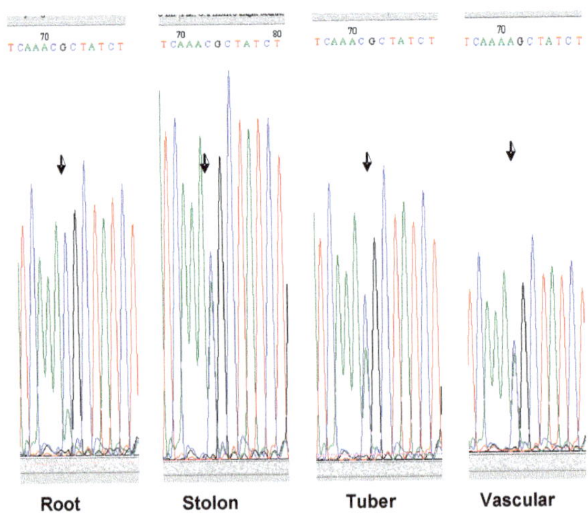

Fig. 7. Part of the raw cDNA sequence files of the C/A SNP region of the abundantly expressed variants (alleles) at the acid invertase locus (primer CA548B) sequenced directly from RT-PCR products. Potato genotype; Shepody. Note that in the cDNA sequence from vascular tissues (the right panel), the amplicon at the SNP read as the A allele by the sequencer, but actually both the C allele (blue peak at the marked SNP) and A allele accumulated to approximately the same abundances.

transcript abundance difference between the two alleles is much smaller (the right 3 panels in Fig. 7). Since the downstream SNP (T/C) was only expressed minimally, the HRM difference should be mainly from this upstream C/A SNP. In conclusion, the HRM curves reflected well the activity ratio between alleles and showed which allele is more actively expressed for a SNP.

Detection of allelic expression evolution in tubers during cold storage by HRM assay

To assess the time course expression, the transcripts of tubers during cold storage using the AGPase gene locus (primer CC1533A) were measured. The expression activity shifted from a higher melting curve (day 0 and day 2 of storage) to a low melting curve allele (day 13 and day 30 of storage) among tubers of the low sugar content variety Atlantic during the 7°C cold storage (Fig. 8: left panel); whereas there was no such a clear shift in the higher sugar content variety Superior (Fig. 8: right panel). The time course also fit well with results from real time PCR (data not shown) but the Lightscanner could identify allelic variants much easily. Therefore, the high-resolution melting curve assay yields complementary and novel information for potato gene expression and demonstrates the power of the high-resolution melting curve method. Although more research is required to investigate whether the HRM curve shift in these varieties is related to the sugar contents in the tubers, the experiment indicates that HRM assay can monitor allelic expression during cold storage and detect differences among potato varieties in their response to cold storage at the transcript abundance level.

Discussion

Sensitivity and reproducibility are crucial parameters for all analyses of gene expression. The improved reliability of the LightScanner platform for transcript abundance measurement is mainly derived from the use of an improved simple RT-PCR analysis that does not require fluorescently labeled probes. The LightScanner platform only needs a small amount of target solution for the assay (Wittwer *et al.*, 2003; Herrmann *et al.*, 2006). For example, nano-scale aliquots of template in the 10 µl of total PCR reaction with one amplification target per tube was performed in parallel in a 96-well-plate format and generated repeatable results. These advantages will enable high throughput screenings. For a direct comparison of various tissues, duplicates or triplicates from different samples can be assayed simultaneously in the 96-well format, and targeted amplicons were accurately determined and duplicated. The technology is sensitive enough to accurately detect the presence of allelic cDNA variants of a gene from various tissues. Significant differences between the treatments were observed between samples.

In the LightScanner software (Idaho Technology Inc, Salt Lake City, UT), there are a number of convenient methods to view and analyze the data, including normalized melting curves and normalized difference curves. These analyses are to highlight differences between individual

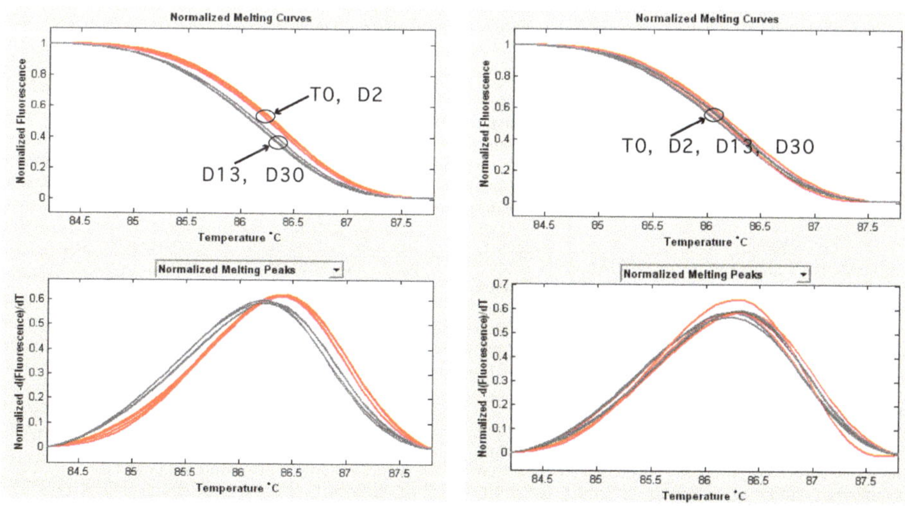

Fig. 8. Expressed dynamics of the ADP glucose pyrophosphorylase gene locus (primer pair CC1533A) gene. Left panel; potato genotype Atlantic. Right panel; potato genotype Superior. Red curves; day 0 (before storage) and day 2 at the start of cold 7°C storage. Grey curves; after 13 days and 30 days of 7°C storage treatments. The HRM assay for the time course study was duplicated. Note that the expression activity shifted to low melting alleles of the SNP during storage of 'Atlantic' tubers; whereas there was no such a shift in 'Superior'.

curves and their baseline. The approach performed here, using unlabeled primers and LC Green, was less expensive than the commonly used fluorescence dyes. Because of its flexibility in experimental design and straightforward way of producing many replications, the direct evidence that the LightScanner platform was sensitive enough to resolve allelic expression differences among various tissues was simple to achieve. The protocols presented here are reliable and easy to handle, compared with various other gene expression analysis methods such as a realtime RT-qPCR method (Nolan *et al.*, 2006), although each protocol has its advantages and disadvantages. This simple platform based on the melting temperature property of cDNA visualizes allelic variants spatially and temporally and assist pursuing each allele's function.

The molecular markers derived from genomic DNA or cDNA derived amplicons and the primers used here segregated in our mapping populations (Li *et al.*, 2005) and can also be directly used for mapping purpose (data not shown). Conventionally, amplicons from different alleles of a known gene cannot be visually differentiated by regular RT-PCR if there is only one or a few nucleotides difference between amplicons. HRM technology largely solved this problem and therefore should be useful for QTL mapping and association genetic analysis with informative, expressed markers, a type of functional allele markers.

The LightScanner HRM system can also use an oligo-probed approach to distinguish two SNPs in the same region amplified by the same pair of primers (Idaho Technology Inc, Salt Lake City, UT). While the oligo-probe approach can give additional useful information in our recent experiments (AGPase, data not shown, unpublished), this non-probe approach is simpler for manipulation and analysis. This simple, fast approach is capable of generating useful information very effectively as demonstrated in this study. The method is so sensitive probably because PCR amplicon regions are very short (150–250 bp) and so closely-linked SNPs behave very similarly, unless a rare genetic recombination occurs within such a short region. This study used a PCR-product sequencing approach to verify the HRM results. However if a database was available for genomic SNPs or cDNA SNPs and their melting temperatures of the SNP regions, the cDNA HRM data with or without the oligo probes could be used as functional allele activity (FAA) markers to sensitively detect the dominantly expressed allele and the transcript abundance ratio between alleles. It is also likely that FAA markers can be converted to functional allele genomic (FAG) markers.

Acknowledgements

We thank S. Whitney for the assistance in using LightScanner, X. Liu for providing some RNA samples, Drs. O. Neyret-Djossou and R. Ritchie for PCR amplicon sequencing, and Dr. D. DeKoeyer for useful discussion and for critical reading of an early version of this manuscript. This research was funded by AAFC A-base and the Canada/NB Embracing Innovation in Agriculture and Agri-Food Program (EIA) through NB Potato Agency (R. Gareau).

References

Alwine, J.C., Kemp, D.J., and Stark, G.R. (1977) Method for detection of specific RNAs in agarose gels by transfer to diazobenzyloxymethyl-paper and hybridization with DNA probes. Proc. Natl. Acad. Sci. U.S.A. *74*, 5350–5354.

Chateigner-Boutin, A.L., and Small, I. (2007) A rapid high-throughput method for the detection and quantification of RNA editing based on high-resolution melting of amplicons. Nucl. Acids Res. *35*, e114.

Flinn, B., Rothwell, C., Griffiths, R., Lague, M., DeKoeyer, D., Sardana, R., Audy, P., Goyer, C., Li, X.Q., Wang-Pruski, G., and Regan, S. (2005) Potato expressed sequence tag generation and analysis using standard and unique cDNA libraries. Plant. Mol. Biol. *59*, 407–433.

Foolad, M.R., and Lin, G.Y (1998) Genetic analysis of low-temperature tolerance during germination in tomato, *Lycopersicon esculentum* Mill. Plant Breed. *117*, 171–176.

Guo, M., Rupe, M.A., Yang, X., Crasta, O., Zinselmeier, C., Smith, O.S., and Bowen, B. (2006) Genome-wide transcript analysis of maize hybrids: allelic additive gene expression and yield heterosis. Theor. Appl. Genet. *113*, 831–845.

Herrmann, M.G., Durtschi, J.D., Bromley, L.K., Wittwer, C.T., and Voelkerding, K.V. (2006) Amplicon DNA melting analysis for mutation scanning and genotyping: Cross-platform comparison of instruments and dyes. Clin. Chem. *52*, 494–503.

Herrmann, M.G., Durtschi, J.D., Wittwer, C.T., and Voelkerding, K.V. (2007) Expanded instrument comparison of amplicon DNA melting analysis for mutation scanning and genotyping. Clin. Chem. *53*, 1544–1548.

Howell, W.M., Jobs, M., and Brookes, A.J. (2002) iFRET: An improved fluorescence system for DNA-melting analysis. Genome Res. *12*, 1401–1407.

Iqbal, M.J., Yaegashi, S., Ahsan, R., Shopinski, K.L., and Lightfoot, D.A. (2005) Root response to *Fusarium solani* f. sp. *glycines*: Temporal accumulation of transcripts in partially resistant and susceptible soybean. Theor. Appl. Genet. *110*, 1429–1438.

Jeong, S., Hahn, Y., Rong, Q., and Pfeifer, K. (2007) Accurate quantitation of allele-specific expression patterns by analysis of DNA melting. Genome Res. *17*, 1093–1100.

Journot-Catalino, H., Somssich, I.E., Roby, D., and Kroj, T. (2006) The transcription factors WRKY11 and WRKY17 act as negative regulators of basal resistance in *Arabidopsis thaliana*. Plant Cell *18*, 3289–3302.

Kugita, M., Yamamoto, Y., Fujikawa, T., Matsumoto, T., and Yoshinaga, K. (2003) RNA editing in hornwort chloroplasts makes more than half the genes functional. Nucl. Acids Res. *31*, 2417–2423.

Li, X.Q., De Jong, H., De Jong, D.M., and De Jong, W.S. (2005) Inheritance and genetic mapping of tuber eye depth in cultivated diploid potatoes. Theoretical and Applied Genetics *110*, 1068–1073.

Meyers, B.C., Galbraith, D.W., Nelson, T., and Agrawal, V. (2004) Methods for transcriptional profiling in plants. Be fruitful and replicate. Plant Physiol. *135*, 637–652.

Montgomery, J., Wittwer, C.T., Palais, R., and Zhou, L. (2007) Simultaneous mutation scanning and genotyping by high-resolution DNA melting analysis. Nat. Protoc. *2*, 59–66.

Nolan, T., Hands, R.E., and Bustin, S.A. (2006) Quantification of mRNA using real-time RT-PCR. Nat. Protoc. *1*, 1559–1582.

SantaLucia Jr, J. (1998) A unified view of polymer, dumbbell, and oligonucleotide DNA nearest-neighbor thermodynamics. Proc. Natl. Acad. Sci. U.S.A. *95*, 1460–1465.

Schenk, P.M., Kazan, K., Wilson, I., Anderson, J.P., Richmond, T., Somerville, S.C., and Manners, J.M. (2000) Coordinated plant defense responses in *Arabidopsis* revealed by microarray analysis. Proc. Natl. Acad. Sci. U.S.A. *97*, 11655–11660.

Syvanen, A.C. (2001) Accessing genetic variation: Genotyping single nucleotide polymorphisms. Nat. Rev. Genet. *2*, 930–942.

Velculescu, V.E., Zhang, L., Vogelstein, B., and Kinzler, K.W. (1995) Serial analysis of gene expression. Science *270*, 484–487.

Wittwer, C.T., Reed, G.H., Gundry, C.N., Vandersteen, J.G., and Pryor, R.J. (2003) High-resolution genotyping by amplicon melting analysis using LCGreen. Clin. Chem. *49*, 853–860.

Zhao, B., Ardales, E.Y., Raymundo, A., Bai, J., Trick, H.N., Leach, J.E., and Hulbert, S.H. (2004) The *avrRxo1* gene from the rice pathogen *Xanthomonas oryzae* pv. *oryzicola* confers a nonhost defense reaction on maize with resistance gene *Rxo1*. Mol. Plant Microbe Interact. *17*, 771–779.

Chapter 2

Multigeneic QTL: The Laccase Encoded within the Soybean Rfs2/rhg1 Locus Inferred to Underlie Part of the Dual Resistance to Cyst Nematode and Sudden Death Syndrome

M.J. Iqbal, R. Ahsan, A.J. Afzal, A. Jamai, K. Meksem , Hany A. El- Shemy, and David A. Lightfoot

Abstract

Multigeneic QTL present significant problems to analysis. Resistance to soybean (*Glycine max* (L) Merr.) sudden death syndrome (SDS) caused by *Fusarium virguliforme* was partly underlain by *QRfs2* that was clustered with, or pleiotropic to, the multigeneic *rhg1* locus providing resistance to soybean cyst nematode (SCN; *Heterodera glycines*). A group of five genes were found between the two markers that delimited the *Rfs2/rhg1* locus. One of the five genes was predicted to encode an unusual diphenol oxidase (laccase; EC 1.10.3.2). The aim of this study was to characterize this member of the soybean laccase gene-family and explore its involvement in SDS resistance. A genomic clone and a full length cDNA was isolated from resistant cultivar 'Forrest' that were different among susceptible cultivars 'Asgrow 3244' and 'Williams 82' at four residues R/H168, I/M271, R/H330, E/K470. Additional differences were found in six of the seven introns and the promoter region. Transcript abundance (TA) among genotypes that varied for resistance to SDS or SCN did not differ significantly. Therefore the protein activity was inferred to underlie resistance. Protein expressed in yeast pYES2/NTB had weak enzyme activity with common substrates but good activity with root phenolics. The Forrest isoform may underlie both *QRfs2* and *rhg1*.

Introduction

Genetic studies have mapped a number of quantitative trait loci (QTL) conferring resistance to sudden death syndrome (SDS; Q*Rfs*, Q*Rsfs1*, Q*Rfs2*, Q*Rfs3*) and soybean cyst nematode (SCN; *rhg1*) on soybean linkage group G (LG G; chromosome 18; Iqbal *et al.*, 2001; Triwitayakorn *et al.*, 2005). The four QTL for resistance to SDS on LG G range from 0.0 to 30 ± 2.5 cM from *rhg*1, a major gene for resistance to SCN (Meksem *et al.*, 1999, 2001). Q*Rfs*2 is about 0.0–0.2 cM from *rhg*1 and may be pleiotropic effect of a single gene (Triwitayakorn *et al.*, 2005; Ruben *et al.*, 2006), Q*Rfs2* reduces the leaf-scorch index, a measure of foliar symptoms that result from the toxins produced by *Fusarium virguliforme*, the causative agent of SDS, but not the root infection severity (percentage of roots infected with *F. virguliforme*). The *Rfs2/rhg*1 region on LG G encompassed between SIUC-Sac13 (Ruben *et al.*, 2006) and AFLP marker ATG4 (Meksem *et al.*, 2001) was shown to contain five genes; a candidate receptor like kinase gene potentially involved in extracellular signal reception and intracellular signal transduction; and four genes encoding enzymes that might be involved in metabolism. Each or all of these genes might underlie resistance to *F. virguliforme* and or SCN (Triwitayakorn *et al.*, 2005; Ruben *et al.*, 2006).

Extensive soybean genome sequence from the susceptible cultivar 'Asgrow 3244' has been released to GenBank encompassing *Rfs2/rhg1* (Hague *et al.*, 2001). In addition significant sequence resources for this region are available from resistant cultivar 'Forrest' (Ruben *et al.*, 2006; Shultz *et al.*, 2006a) and susceptible culivar 'Williams 82' (J. Schmutz personal communication 2006). Extensive sequencing of the receptor like kinase in 32 cultivars and 112 plant introductions (PIs) showed there were 9 alleles encoding 5 different proteins (Ruben *et al.*, 2006). Allele 1 was perfectly associated with resistance to SCN HgType 0 (race 3). Allele 1 has also been shown to be associated with SDS resistance across a very wide collection of germplasm whereas allele 2 was more associated with susceptibility to SDS (Gibson, 1994; Njiti *et al.*, 1997; 2002). The allelic diversity of the neighboring genes has not been well characterized to date. However, since linkage disequilibrium in soybean is often large (97–536 kbp; Hyten *et al.*, 2007) significant numbers of alleles among the linked genes are expected.

Diphenol oxidase laccase (hereafter called as laccase) was considered a strong candidate for Q*Rfs*2 (Triwitayakorn, 2005) and may be part of *rhg1* (Ruben *et al.*, 2006). The laccase enzyme (p-diphenol:O2 oxidoreductase; EC 1.10.3.2) is a blue copper-containing oxidase found in plants, fungi, bacteria (Diamantidis *et al.*, 2000) and arthropods (Thomas *et al.*, 1989; Cardenas and Dankert, 2000). Laccases in plants are present as large multigene families (18–20 members) that can be classified into 6 major sub-groups some of which predate the monocot, dicot split (McCaig *et al.*, 2005). Phenotypes of four mutants in Arabidopsis include seed coat color, root development, flowering time though the majority (8) showed no phenotype (Cai *et al.*, 2006). Many members are expressed in roots. Together this data suggests it is unlikely most laccases participate directly in cell wall lignifications although a few may (Ranocha *et al.*, 2002). Other roles include the hydroxlation of flavonoids, formation of proanthocyanidin or tannin and polymerization of phenolic compounds which protect plants from pathogen and insect attack. A variant laccase might detoxify phenolic fungal toxins, like monorden (Baker and Nemec, 1994) or reduce the frequency of SCN feeding site development. Alternately a laccase expressed in the roots to increase cell wall lignification (Lozovaya *et al.*, 2004) might protect against fungal ingress or spread and therefore decrease the amount of toxin produced or translocated (Lozovaya *et al.*, 2005).

Laccase genes have been found in all major seed plants and have been well characterized in *Arabidopsis thaliana* (Cai *et al.*, 2006), ryegrass (*L. perenne*; Gavnholt *et al.*, 2002), maple (*Acer pseudoplatanus*; La Fayette *et al.*, 1999), tobacco (*Nicotiana tabacum*; Kiefer-Myer *et al.*, 1996), poplar (*Populus trichocarpa*; Ranocha *et al.*, 1999) and yellow poplar (*Liriodendron tulipifera*; Fayette *et al.*, 1999). Several Poplar laccase mRNAs were expressed in stems, but not in leaves and roots (Ranocha *et al.*, 1999). Ryegrass laccase mRNA was differentially expressed in stem and meristem (Gavnholt *et al.*, 2002). However, laccase gene family sizes are large (18–20 members) in most species and expression of at least some family member is found in all organs (Cai *et al.*, 2006).

Based on the presence of laccase within the region encompassing the *Rfs2/rhg1* resistant locus and the multifunctional role of the enzyme class, it was hypothesized that the soybean laccase might be involved in resistance to SDS and SCN. This paper reports the characterization of the laccase enzyme; identification of laccase alleles; analysis of transcript abundances in roots of several soybean varieties varying in partial resistance to SDS and SCN; and associations with the alleles at the receptor like kinase within the *rhg1* region The possibility that laccase is a candidate for one of the genes underlying both the *rhg1* and the Q*Rfs*2 locus is discussed.

Materials and methods

Plant material

Soybean varieties and genotypes used in this study are listed in Table 1. Alleles for *rhg1* are as listed in Ruben *et al.* (2006) from receptor like kinase sequences. Recombinant inbred line (RIL) ExF23 contains the favorable alleles of the 6 SDS QTL and considered as partially resistant. RIL ExF85 contains the susceptible alleles of the 6 SDS QTL and was considered as susceptible (Iqbal *et al.*, 2001; Njiti *et al.*, 2001). Forrest contains 4 beneficial QTL and Essex two. The complement of segregating QTL were not completely equivalent to Essex and Forrest in 'Flyer', 'Hartwig', 'Ripley' and 'Spencer' (Farias-Neto *et al.*, 2007; Kazi *et al.*, 2008) or were not yet characterized ('Hamilton', 'Jack').

Table 1. Soybean varieties and recombinant inbred lines (RIL) studied for the TA of laccase diphenol oxidase like sequence (AY113187) in roots inoculated with SDS pathogen at 7 days.

			Change in TA			
	Name	SDS[a]	PR	S	*rhg1* allele[b]	Seed source
1	RIL23	PR	1.48		1	Southern Illinois University
2	RIL85	S		2.195	6	Southern Illinois University
3	Jack	PR	0.546		2	USDA NSRC, Urbana, IL
4	Flyer	S		0.56	4	USDA NSRC, Urbana, IL
5	Ripley	PR	0.928		5	USDA NSRC, Urbana, IL
6	Hamilton	PR	1.094		1	USDA NSRC, Urbana, IL
7	Spencer	S		1.000	4	USDA NSRC, Urbana, IL
8	Hartwig	PR	0.57		1	USDA NSRC, Urbana, IL
9	Essex	S		nd	6	Southern Illinois University
10	Forrest	PR	nd		1	Southern Illinois University
11	Williams82	S		nd	6	USDA NSRC, Urbana, IL
12	A3244	S		nd	4	USDA NSRC, Urbana, IL
Mean± SEM[c]			0.92±0.17	1.25±0.42		

[a] The resistance (R) and susceptibility (S) is based on Njiti *et al.* (2001).

[b] The *rhg1* allele is after Ruben *et al.*, (2006) based on the receptor like kinase haplotype.

[c] The two means are not significantly different (p >0.05).

SDS assays

Seeds were germinated in sterilized sand:soil (1:1) mix in a growth chamber and inoculated with *F. virguliforme* spores as described earlier (Iqbal *et al.*, 2002, 2005). Root samples from control and inoculated plants were collected at seventh day after inoculation.

RNA isolation and cDNA synthesis

Total RNA from frozen roots or leaves was extracted using plant RNeasy Mini Kit (QIAGEN, Valencia, CA, USA) according to the manufacturer's instructions. Contaminating DNAs were removed by DNAase treatment during the RNA isolation process. For the isolation of full length laccase, cDNA was synthesized using SMART™ RACE cDNA Amplification Kit (BD Biosciences, Palo Alto, CA, USA) according to manufacturer's instructions. The cDNA was amplified using laccase gene (AY113187) specific primers (Forward 5'ATGGAGCCTGCCAAAACCATTCAC3'; Reverse 5'CTAACAAAGAGGAAGATCCACAGGA3'). The PCR product was cloned in to pGEM-T vector (Promega, Madison, WI, USA) and transformed according to the manufacturer's instructions.

Expression of laccase in yeast

The cloned cDNA was removed from vector by *Eco*R1 digestion; gel purified and sub-cloned into *Eco*R1-digested pYES2/NTB (Invitrogen, Carlsbad, CA) fusion vector according to the manufacturer instructions. The presence of inserts was confirmed by restriction analysis followed by gel electrophoresis. The orientation of the insert was determined by DNA sequencing using ABI 377 automated DNA sequencer.

Total RNA was extracted from control INVSc1 yeast cells, non-induced and induced samples using plant RNeasy Mini Kit (QIAGEN, Valencia, CA) and the expression of cloned laccase was confirmed by Northen hybridization using NorthernMax™ kit (Ambion, Austin, TX) according to their instructions.

Total proteins from induced, non-induced and control yeast cells and the recombinant expressed protein were detected by Western hybridization using Anti-Xpress™ (Cat # R910–25, Invitrogen, Carlsbad, CA) as primary antibody and antimouse-peroxidase conjugate (Cat # NA931VS, Amersham, Piscataway, NJ) as secondary antibody. A second Western was probed solely with a mouse monoclonal IgG antibody Anti-Xpress™ HRP (Cat # R911–25, Invitrogen, Carlsbad, CA) according to the manufacturer's instructions.

Enzyme assays

Laccase activity was calculated from the rate of oxidation of 5 mM of 2, 2'-azinobis-(3-ethylbenzothiazoline-6-sulfonic acid) (ABTS) in 100 mM sodium acetate (pH 5.0) at 420 nm after Hoopes and Dean (2001) and the oxidation of O-phenylenediamine (OPDA) after Zuyun *et al.*, (1998). In gel staining following electrophoresis was in 50 mM sodium acetate (pH 5.0) containing 1% (v/v) dimethyl sulfoxide and 2 mM 1,8-diaminonaphthalene and incubated at 30°C until activity bands developed (Hoopes and Dean, 2001).

Analysis of laccase transcript abundance in response to F. virguliforme *inoculation*

Initial laccase gene-family transcript abundance (TA) measurements used reverse Northern hybridization (Iqbal *et al.*, 2005) in *F. virguliforme* inoculated and non-inoculated roots of RIL 23 and 'Essex'. An EST representing *G. max* laccase (AI495260) was spotted on a membrane. Probe synthesis, hybridization and post-hybridization treatments and data analysis were carried out as described in Iqbal *et al.* (2002, 2005).

For RT-PCR, total RNA was isolated from inoculated and control roots. cDNA was synthesized from 1 µg using iScript™ cDNA Synthesis Kit (BIORAD, Hercules, CA). Quantitative real time PCR (RT-PCR) amplification was carried out in a iQ™ 5 Multicolor Real Time PCR Detection System (BIORAD, Hercules, CA) using iQ™ SYBR® Green Supermix (BIORAD, Hercules, CA) and 10 pmoles of forward and reverse primers in 20 µl reaction volume. The transcript abundances (TA) of two different laccases were estimated. Primers selective for AY113187, the laccase at the *Rfs2/rhg1* locus, (Scaffold 121, LgG, near Satt309) were made. The forward primer was 5'GTCCATCTTGCAGGCTCACCAC3' and the reverse primer was 5'TTGGGTCGAAGTTCCC GAAACC3'). Primers selective for the EST AI495260 (forward primer 5' ATGCATTGCCACTTTGATGTCC 3' and reverse primer 5' AGTAGGCAAACCAAAATTCCGG 3') was measured. This laccase is potentially encoded at two loci in the genome, both on LG H, Scaffold_112 near Satt568 and Scaffold 137, near Satt247. The reactions were carried out in triplicate. Soybean β-tubulin (forward primer 5' CGCACCCTTAAGCTCACCACC 3', reverse primer 5' TTTCCACGGACATCGCAGAAGC 3') was amplified in duplicates from the same cDNA as a control in separate reactions to normalize the data. Data normalization and fold changes in TA between treated and non-treated samples were calculated by $2^{-\Delta\Delta Ct}$ method (Livak and Schmittgen, 2001). Absolute values for TA among cultivars were calculated by reference to the fresh weight of roots from which RNA was extracted and compared to the 18S rRNA band fluorescence and the tubulin amplicon abundance (Iqbal *et al.*, 2005).

Determination of laccase copy number in the soybean genome

A minimum tiling path (MTP, build 2) of soybean bacterial artificial chromosome library (BAC) developed at Southern Illinois University, Carbondale, IL (Shultz *et al.*, 2006b) contained 8, 064 clones representing ~1-fold coverage of the soybean genome. In order to determine laccase homologous sequences, a $\alpha^{32}P$ dCTP labeled laccase cDNA probe was hybridized to this set of MTP clones as described earlier (Shopinski *et al.*, 2006) and number of hybridizing BACs were counted to estimate the number of paralogs in the genome.

Results and discussion

Sequence analysis of alleles of the laccase at Rfs2/rhg1.

Alignment of the 1, 770 bp protein coding portion of the cDNA (AY113187) with 4, 615 bp from the equivalent genomic sequence (AF527604) of the Forrest allele of the laccase at *Rfs2/rhg1* showed the gene was encoded by seven exons (Fig. 1; 47–172, 286–437, 1036–1280, 1537–1665, 2560–3101, 3654–4098, 4452–4582 bp). The protein predicted to be encoded had a theoretical MW of 64.934 MDa and a pI of 9.23 from 590 amino acids. Two of the introns (I2 and I4) were unusually large at 599 and 895 bp.

Comparison of the Forrest to the Asgrow 3244 allele of the laccase at *Rfs2/rhg1* showed 4 conservative changes in amino acid sequence; R/H168, I/M 271, R/H 330, E/K 470 (Fig. 1). Non-conservative changes from residues 112 to 116 and S/A 356 were frame shift errors in the Forrest allele sequences posted in 2002. Comparing genomic sequences showed no SNPs in the second or third introns but a 70–80 bp region of divergence in the first intron of the Forrest allele; a 159 bp region in the fourth intron; a 35 bp region in the fifth intron; and a 92 bp region in the sixth intron.

Comparison of the Williams 82 allele of the laccase at *Rfs2/rhg1* to Forrest and Asgrow 3244 showed equally significant similarity with multiple SNPs (9–24) in every intron that distinguished the 3 alleles. In the first intron Williams 82 was very different from both Forrest and A3244 showing just 85 % identical sequence due to the 70–80 bp insertion (Supplemental Fig. 1).

Comparing promoter regions showed significant differences among the three alleles that may have effects on transcript abundance. The promoter of Forrest shared just 89 % identity in the first 300 bp that encompassed the core promoter with the Williams 82 and A3244 alleles that did not differ from one another in that region. Differences encompassed but did not disrupt the potential TATA box regions at −40, −32 and −11. From −300 bp to −2, 000 bp distal to the transcription start site alleles were 100 % identical between Forrest, Williams 82 and Asgrow 3244. In summary, the sufficient differences found among the three alleles suggested that transcript abundance might differ significantly between cultivars with different alleles, particularly the Forrest allele.

Effect of F. virguliforme *infestation on laccase mRNA abundance*

Reverse Northern hybridizations indicated some correlation between laccase TA and resistance (Fig. 2) at later stages (day 10) of root-pathogen interaction. The initial decrease in TA at day 1 after inoculation looks most likely the result of stress caused by the process of inoculation and transfer of roots into new media. The TA at day 2 and 7 after inoculation does not indicate any significant change in the resistant and susceptible inoculated roots. Days 1–7 are considered as early to middle phases of the *F. virguliforme* and soybean root interaction (Iqbal *et al.*, 2005). However the probe used was not strictly gene specific and may have measured the TA of a set of related laccase genes in the soybean genome (Fig. 3). Further the day 10 effect may not be specific because nearly all transcript abundances are reduced by this stage in a susceptible cultivar (Iqbal *et al.*, 2005).

The inoculation and TA analysis experiment were conducted multiple times in ExF23 and ExF85 RILs contrasting for the SDS QTL concentrating on day

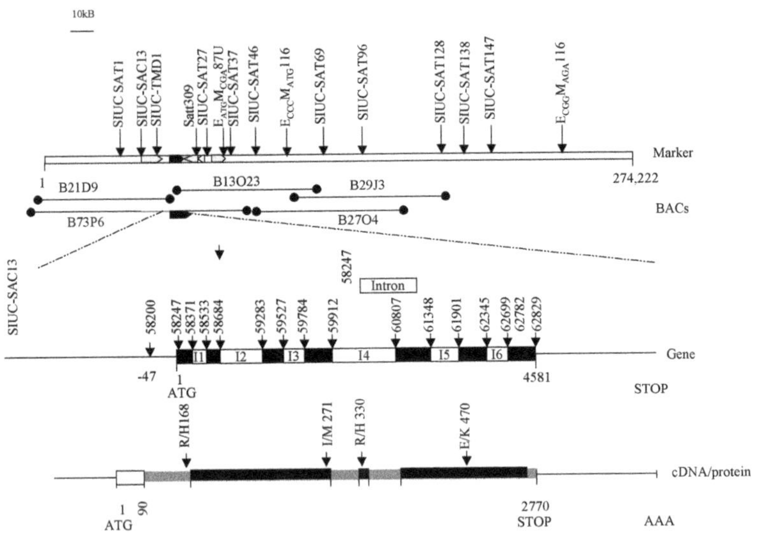

Fig. 1. The physical map of the laccase gene found at the *rhg1* locus. Shown by the open white bar are marker positions in relation to the first 274, 222 kbp of AX196295. An interval spanning 10 Kbp encoding the putative *laccase* gene is shown by a black arrow. In the middle panel the introns (white boxes) and exons (black boxes) are shown. In the lower panel the positions of markers and amino acid substitutions in the cDNA among the alleles of Forrest, Asgrow 3244 and Williams 82 are shown. Copper binding domains are indicated by the black boxes.

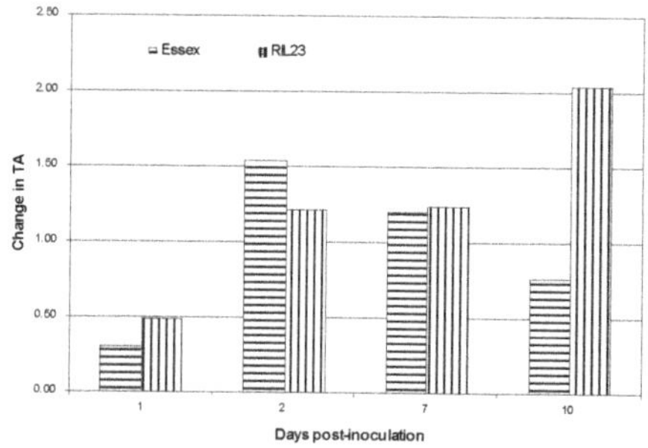

Fig. 2. Change in transcript abundance (TA) of AY113187, the laccase at the *Rfs2/rhg1* locus, measured by reverse Northern hybridization in roots of RIL23 and Essex at 1–10 dai. The hybridization intensity measured from the inoculated samples was divided by that of non-inoculated samples. Scale of 1 on Y-axis represents no change in TA.

7 after inoculation when many gene transcript differ between resistant and susceptible genotypes but not because transcription has ceased in the susceptible cultivars (Iqbal *et al.*, 2005). The results indicated some significant differences among cultivars. However, there was no significant correlation between change in TA of laccase and the SDS response. In contrast, the number of favorable alleles of SDS QTL (Table 1; Fig. 4) and the allele at *Rfs2/rhg1* predicted RLK were associated with SDS response. Together the results indicate that there is no direct correlation between inoculation induced changes in laccase transcript abundance at 7 days after inoculation (dai) and the heritable resistance of a genotype to SDS.

In fact the laccase at *Rfs2/rhg1* was increased in TA in roots at day 10 after inoculation in RIL23 (Figs 4 & 5). However, the roots of the susceptible genotypes were dead to the extent that there were not many intact transcripts. The laccase TA increased after inoculation only in RIL23 but not Hartwig or Jack that also show root resistance. However, it is possible the assay used had an effect because the correlation between field resistance to SDS and the seedling assay used here is not perfect (Njiti *et al.*, 2001). The expression of laccase could be increased at later developmental stages of the soybean roots (Njiti, 1997; Luo *et al.*, 1999). Taken together though the data indicate it is the steady state amount of laccase mRNA and/or the isoform expressed from it that are most likely underlying the effects on resistance to SDS among different cultivars.

Identification of diphenol oxidase (laccase) sequences in soybean genome

The lack of correlation between the inherited resistance and the laccase expression raised questions as to the effect of the gene on overall laccase gene family TA. Therefore, the copy number of nearly identical laccase paralogs present in the soybean genome was investigated by high stringency hybridizations to a set of BAC clones covering a minimum tiling path (MTP) of the soybean genome (Shultz *et al.*, 2006b). There were 19 BAC clones that hybridized consistently on two filters; B37O05, B38G01, B40D11, B44J09, B10A18, B48I14, B44I16, H38F23, H25M14, H42H20, H57E15, H52G24, H07F23, H64G11, H72M11, H62O22, H71F11, H69G07 and H76L13. The BAC clones B44J09 and B48I14 were present on soybean LG A1 (from Build 4 and 3), H62O22 was identified on soybean LG A2 (Build 3), H64G11 and H69G07 on soybean LG D1BW (Build 3) and H07F23 was identified on soybean LG O. BAC clone B44J09, positive for laccase was identified on soybean LG J. in the version 4 of the soybean physical map. The rest of the BAC clones could not yet be assigned to specific linkage groups but were from different contigs. The soybean BAC clones B21D9 and B73P6 were not present on the membrane because they were not fingerprinted (Wu *et al.*, 2004). The presence of at least 20 copies (19 plus the one at the *Rfs2/rhg1* locus) of laccase-like sequences in the soybean genome was not surprising. In ryegrass (*Lolium perenne*), Gavnholt *et al.* (2002) reported the presence of 25 laccase genes. The Arabidopsis database search identified 12 predicted laccase genes (Cai *et al.*, 2006). The difference in the number of laccase genes in *Arabidopsis* and ryegrass was attributed to the differences in their genome size. If that is true, then the presence of 20 laccase genes in soybean can be attributed to its genome size and the polyploid nature of parts of its

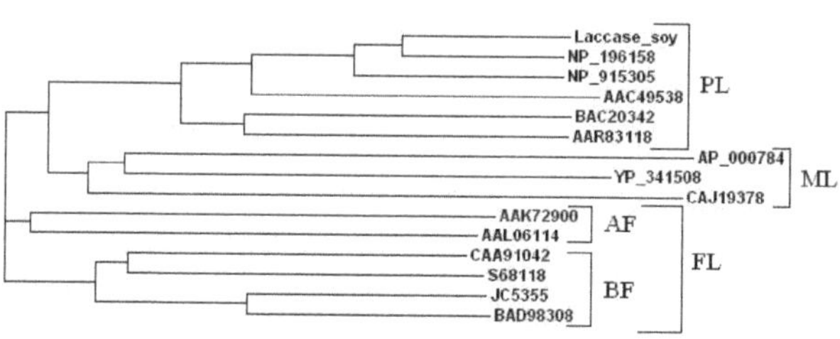

Fig. 3. Amino acid similarity between plant (PL), microbial (ML) and fungal (FL) laccases. The plant sequences included were of soybean (Laccase_soy; AY113187, the laccase at the *Rfs2/rhg1* locus), *Arabidopsis thaliana* (NP_196158), *Oryza sativa* (NP_915305), Tobacco (AAC49538), *Rhus vernicifera* (BAC20342) and *Gossypium arboreum* (AAR83118). The microbial laccases include *Escherichia coli* (AP_000784), *Pseudoalteromonas haloplanktis* (YP_341508) and *Pedomicrobium* sp. (CAJ19378). The fungal laccases include two ascomycetes fungi (AF) *Fusarium proliferatum* (AAK72900) *Botryotinia fuckeliana* (AAL06114) and four basidiomycetes fungi (BF) *Thanatephorus cucumeris* (CAA91042), *Rhizoctonia solani* (S68118), *Trametes villosa* (JC5355) and *Trametes versicolor* (BAD98308. The tree is based on Multiple Alignments using ClustalW v1.8.

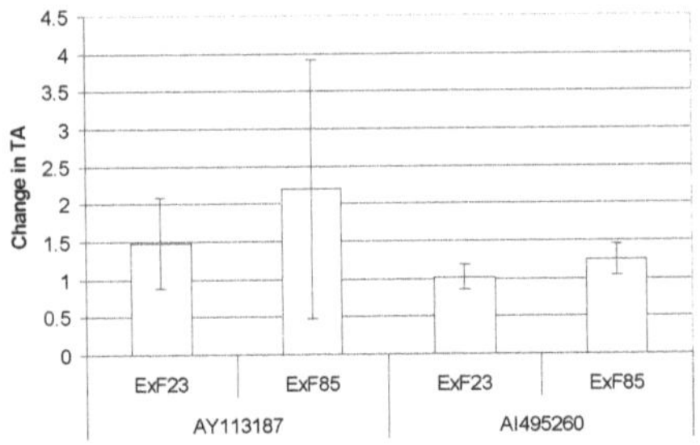

Fig. 4. Change in TA of two ESTs (AY113187 the laccase at the *Rfs2/rhg1* locus and AI495260 a duplicated laccase on LgH and) representing two soybean laccases in RIL23 and RIL85 contrasting for the presence of SDS QTL at 7 dai. The results are mean of 3 independent experiments. In each experiment, roots of 3–5 plants were pooled together and TA was measured by quantitative real time-PCR. Bars indicated standard deviation.

genome (Shultz *et al.*, 2006a). However, whether all the soybean laccase like sequences identified are functional genes has not been determined.

Orthologs of the laccase at Rfs2/rhg1

Plant laccases are closely related. The deduced amino acid sequence clusters the plant laccases form are distinct (Fig. 3) from microbial and fungal laccases. That separation reflects genetic similarities, evolutionary distances and similar but distinct physiological roles in each species. There is no evolutionary distinction between monocot, dicot and angiosperm laccases (Gavnholt *et al.*, 2002). However, a high rate of divergence among dicot laccases has been reported (Gavnholt *et al.*, 2002). The differences in amino acid sequence between the plant laccases, fungal laccases and ascorbate oxidases are less pronounced (Ranocha *et al.*, 1999), but enough to form separate groups. Within plants, the soybean laccase paralog analyzed here showed a ~75% identity with an *Arabidopsis thaliana* laccase (AtLAC12; Cai *et al.*, 2006), 67% identity with an *Oryza sativa* (cv. japonica) putative laccase and 63% identity with a *Pinus taeda* laccase. Mutants in AtLAC 12 did not show any phenotypes (Cai *et al.*, 2006).

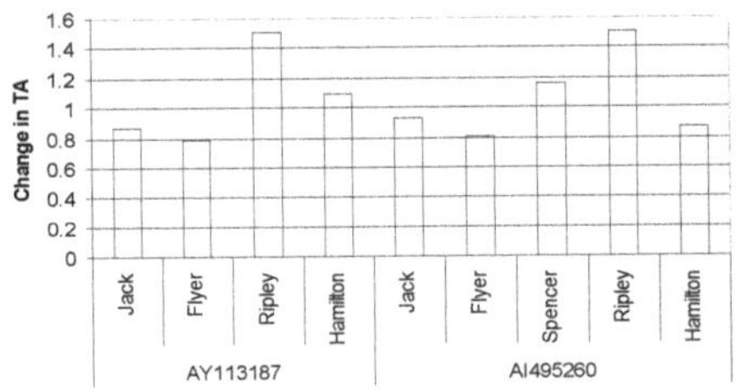

Fig. 5. Change in TA of two ESTs AY113187 (at *Rfs2/rhg1* on LgG) and AI495260 (on LgH) representing two soybean laccases in different soybean varieties at 7 dai. The roots of 3–5 plants were pooled together and TA was measured by quantitative real time-PCR. Each experiment was triplicated (three technical replicates) and means are plotted.

Characterization of the soybean laccase enzyme

The soybean laccase cDNA representing the protein coding portion of the Forrest allele was placed in a yeast expression vector. The expression vector clone was re-sequenced from both ends and no changes were found. The expression of the laccase in yeast was confirmed first by Northern hybridization (Fig. 6, A and B). There was a strong hybridization signal in the lanes containing the RNA samples from galactose-induced yeast cells. However, there was a weak signal in the RNA samples that were isolated from cells that were not induced indicating low constitutive transcription abundance. Yeast cells that were not transformed by the recombinant INVSc1 carrying the laccase gene did not show any expression (Fig. 6B; lane 5).

Western hybridization showed that the Anti-Xpress™ antibody reacted with the induced fusion protein containing the Xpress™ epitope (Fig. 6c, Lane 3–7). No signal was detected from non-induced cells (Lane 8). A positive control, lacZ containing Xpress™ epitope for the detection of lacZ was also used (Lane 9). The lacZ gene encoding β-galactosidase was expressed in yeast cell under the control of the GAL1 promoter. The N-terminal encoding Xpress™ epitope and polyhistidine tag will add ~3.4 kDa to the 67 kDa of recombinant protein. The expressed protein was found to be 70.3 kDa. Therefore the actual size of the diphenol oxidase laccase protein was 67 kDa, confirming the size of the protein as identified from the

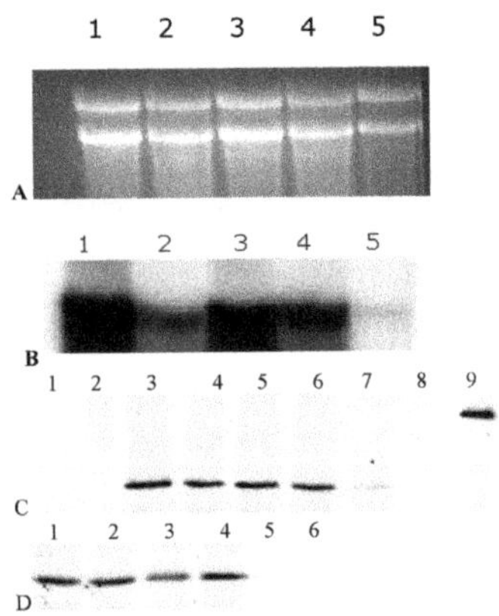

Fig. 6. Expression of diphenol oxidase laccase from AY113187, the laccase at the *Rfs2/rhg1* locus, in induced and non-induced yeast samples by Northern hybridization and Western hybridization. Panel A: Total RNA. Panel B: Northern hybridization. In each lanes 1 and 3 were induced, lanes 2 and 4 were non-induced, lane 5 contained RNA isolated from INSc1 yeast cells. PanelC shows Western hybridization for the detection of expression of the recombinant fusion protein using Anti-Xpress™ antibody. Lanes 1 and 8 contained proteins from non-induced cells; lane 2 was a protein marker, lanes 3 to 7 contained proteins from induced cells; and lane 9 contained tagged lacZ protein. Panel D shows Western hybridization with the Anti-Xpress™-HRP (horseradish peroxidase-conjugated) antibody that react with induced fusion protein containing Xpress™ epitope. Lanes 1 to 4 contained proteins from recombinant yeast cells induced for the expression of diphenol oxidase laccase; lanes 5 and 6 contained proteins from non-induced cells.

deduced amino acid sequence. A number of common substrates were used to determine the laccase activity of the expressed protein. However, very low level of activity was observed in the gel assay (results not shown). The level of activity was very low and could not be quantified by either flourimetric assay. Phenol oxidases usually have specificity toward particular electron acceptors but be able to use a wide variety of electron donors (Zuyun *et al.*, 1998; Hoopes and Dean, 2001). The soybean laccase described here appeared to be specific for both. Alternately the protein produced in yeast was not properly folded or lacked an essential cofactor in yeast.

Conclusions

Map based cloning is an important tool in soybean gene identification (Searle *et al.*, 2003; Ashfield *et al.*, 2003; Gao *et al.*, 2005). However, the association of metabolic genes present in the QTL regions with traits is problematic (Hobbs *et al.*, 2004). The clustering of genes and iterations among enzyme encoding genes might each only have small effects on the trait. SDS resistance has two components, resistance to root infection and resistance to the leaf scorch caused by the toxin (Njiti *et al.*, 1997; 1998). The increases in laccase TA after 10 dai in *F. virguliforme* infested roots in a partially resistant RIL23 line may be due to the ability of the other components of resistance in that genotype to increase the abundance of defense related transcripts. Laccase TA might have only a supporting role in partial resistance. Based on the TA abundance changes in the inoculated roots, it is not likely that increases in the expression of laccase contributes to the SDS QTL named *Rfs2* or the cyst nematode resitance locus *rhg1* (Afzal, 2007; Afzal *et al.*, 2009). Rather the higher expression found in resistant cultivars and the amino acid changes in the enzyme isoform found in resistant cultivars were inferred to underlie resistance. Further analysis will require analysis of this locus on linkage group G and the syntenic locus on what appears to be linkage group O BAC clone H07F23 with gene specific probes, transgenic plants and allelic mutant series.

Acknowledgements

This research was funded over the past 11 years in part by grants from the NSF 9872635, and USB 2228–6228. The integrated genetic and physical map was based upon work supported by the National Science Foundation under Grant No. 9872635. Any opinions, findings, and conclusions or recommendations expressed in this material are those of the author(s) and do not necessarily reflect the views of the National Science Foundation. The past and continuing support of SIUC, Office of the Vice Chancellor for Research to MJI, RA, JA and DAL is appreciated. The authors thank Dr. P. Gibson, O Myers Jr., and M. Schmidt for assistance with germplasm development and maintenance from 1991–2000 and Chris Town (TIGR) for assistance with DNA sequence analysis and interpretation. We thank the "Soybean Genome Project, at DoE Joint Genome Institute" for release of the WGS reads and scaffolds.

References

Afzal, A.J., Natarajan, A., Saini, N., Iqbal, M.J., Geisler, M., El Shemy, H., Mungur, R., Wilmitzer, L., and Lightfoot, D.A. (2009) The nematode resistance allele at the *rhg1* locus alters the proteome and metabolome of soybean roots. Plant Physiol. (in press).

Ashfield, T., Bocian, A., Held, D., Henk, A., Marek, L., Danesh, D., Penuela, S., Meksem, K., and Lightfoot, D.A., Young, N., Shoemaker, R., and Innes, R. (2003) Genetic and physical mapping of the soybean *Rpg1-b* disease resistance gene reveals a complex locus containing several tightly linked families of NBS/LRR genes. Mol. Plant Microbe Interact. *16*, 817–826.

Baker, R.A., and Nemec, S. (1994) Soybean sudden death syndrome, Isolation and identification of a new phytotoxin from cultures of the causal agent, *Fusarium solani* (abstract) Phytopathol. *84*, 1144.

Cai, X., Davis, E.J., Ballif, J., Liang, M., Bushman, E., Haroldsen, V., Torabinejad, J., and Wu, Y. (2006) Mutant identification and characterization of the laccase gene family in Arabidopsis. J. Exp. Bot. *57*, 2563–2569.

Cardenas W, and Dankert, J.R. (2000) Cresolase, catecholase and laccase activities in haemocytes of the red swamp crayfish. Fish Shellfish Immunol. *10*, 33–46.

Davin, L.B., Bedgar, D.L., Katayama, T., and Lewis, N.G. (1992) On the stereoselective synthesis of (+)-pinoresinol in *Forsythia suspense* from its achiral precursor, coniferyl alcohol Phytochemistry *31*, 3869–3874.

Dean, J.F.D., and Eriksson, K.-E.L. (1994) Laccase and deposition of lignin in vascular plants. Holzforschung *48*, 21–33.

Diammantidis, G., Effosse, A., Potier, P., and Bally, R. (2000) Purification and characterization of the first bacterial laccase in the rhizospheric bacterium *Azospirllum lipoferum*. Soil Biol. Biochem. *342*, 919–927.

Farias-Neto, A.F., Hashmi, R., Schmidt, M.E., Carlson, S.R., Hartman, G.L., Li, S., Nelson, R.L., and Diers, B.W. (2007) Mapping and confirmation of a sudden death syndrome resistance QTL on linkage group D2 from the soybean genotypes 'PI 567374' and 'Ripley'. Mol. Breed. *20*, 53–62.

Gao, H., Narayanan, N.N., Ellison, L., and Bhattacharyya, M.K. (2005) Two classes of highly similar coiled coil-nucleotide binding-leucine rich repeat genes isolated from the Rps1-k locus encode Phytophthora resistance in soybean. Mol. Plant Microbe Interact. *18*, 1035–1045.

Gavnholt, B., Larsen, K., and Rasmussen, S.K. (2002) Isolation and characterization of laccase cDNAs from meristematic and stem tissue of ryegrass (*Lolium perenne*) Plant Sci. *162*, 581–590.

Gibson, P.T., Shenaut, M.A., Njiti, V.N., Suttner, R.J., and Myers Jr, O. (1994) Soybean varietal response to sudden death syndrome. In, D. Wilkinson (ed), Proc Twenty-fourth Soybean Seed Res. Conf., Chicago, IL 6–7 Dec, 1994 (Washington DC: Am Seed Trade Assoc), pp 436–446.

Hauge, B.M., Wang, M.L., Parsons, J.D., and Parnell, L.D. (2001) Nucleic acid molecules and other molecules associated with soybean cyst nematode resistance. WO 01/51627 PCT/US01/00552 Patent # 20030005491.

Hobbs, D.H., Flintham, J.E., and Hills, M.J. (2004) Genetic control of storage oil synthesis in seeds of Arabidopsis. Plant Physiol. *136*, 3341–3349.

Hoopes, J.T., and Dean, J.F.D. (2001) Staining electrophoretic gels for laccase and peroxidase activity using 1,8-diaminonaphthalene. Anal. Biochem. *293*, 96–101.

Hyten, D.L., Choi, I.Y., Song, Q., Shoemaker, R.C., Nelson, R.L., Costa, J.M., Specht, J.E., and Cregan, P.B. (2007) Highly variable patterns of linkage disequilibrium in multiple soybean populations. Genetics *175*, 1937–1944.

Iqbal, M.J., Meksem, K., Njiti, V.N., Kassem, M.A., and Lightfoot, D.A. (2001) Microsatellite markers identify three additional quantative trait loci for resistance to soybean sudden-death syndrome (SDS) in Essex x Forrest RILs. Theor. Appl. Genet. *102*, 187–192.

Iqbal, M.J., Afzal, A.J., Yaegashi, S., Ruben, E., Triwitayakorn, K., Nijiti, V.N., Ahsan, R., Wood, A.J., and Lightfoot, D.A. (2002) A pyramid of loci for partial resistance to *Fusarium solani* f. sp. *glycines* maintains *Myo*-inositol 1-phosphate synthase expression in soybean roots. Theor. Appl. Genet. *105*, 1115–1123.

Iqbal, M.J., Yaegashi, S., Ahsan, R., Shopinski, K., and Lightfoot, D.A. (2005) Root response to *F. solani* f. sp. *glycines*, Temporal accumulation of transcripts in partially resistant and susceptible soybean. Theor. Appl. Genet. *110*, 1429–1438.

Kazi, S., Shultz, J.L., Afzal, A.J., Njiti, V.N., and Lightfoot, D.A. (2007) Different loci underlie resistance to soybean SDS root infection and leaf scorch in RILs from 'Flyer' by 'Hartwig'. Theor. Appl. Genet. *116*, 967–977.

Kiefer-Myer, M.-C., Gomord, V., O'Connell, A., Halpin, C., and Faye, L. (1996) Cloning and sequence analysis of laccase-encoding cDNA clones from tobacco. Gene *178*, 205–207.

LaFayette, P.R., Eriksson, K.-E.L., and Dean, J.F.D. (1999) Characterization and heterologous expression of laccase cDNAs from xylem tissue of yellow-poplar (*Liriodendron tulipifera*) Plant Mol. Biol. *40*, 23–35.

Livak, K.J., and Schmittgen, T.D. (2001) Analysis of relative gene expression data using real time quantitative PCR and the $2^{-\Delta\Delta Ct}$ method. Methods *24*, 402–408.

Lozovaya, V.V., Lygin, A.V., Zernova, O.V., Li, S., Hartman, G.L., and Widholm, J.M. (2004) Isoflavonoid accumulation in soybean hairy roots upon treatment with *Fusarium solani*. Plant Physiol. Biochem. *42*, 671–679.

Lozovaya, V.V., Lygin, A.V., Zernova, O.V., Li, S., Widholm, J.M., and Hartman, G.L. (2005) Lignin degradation by *Fusarium solani* f. sp. *glycines*. Plant Dis. *90*, 77–82.

Luo, Y., Myers Jr, O., Lightfoot, D.A., and Schmidt, M.E. (1999) Root colonization of soybean cultivars in the field by *Fusarium solani* f. sp. *glycines*. Plant Dis. *83*, 1155–1159.

McCaig, B.C., Meagher, R.B., and Dean, J.F.D. (2005) Gene structure and molecular analysis of the laccase-like multicopper oxidase (LMCO) gene family in *Arabidopsis thaliana*. Planta *221*, 619–636.

Meksem, K., Doubler, T.W., Chancharoenchai, K., Njiti, V.N., Chang, S.J.C., Rao-Arelli, A.P., Cregan, P.E., Gray, L.E., Gibson, P.T., and Lightfoot, D.A. (1999) Clustering among loci underlying soybean resistance to *Fusarium solani*, SDS and SCN in near-isogenic lines. Theor. Appl. Genet. *99*, 1131–1142.

Meksem, K., Pantazopoulos, P., Njiti, V.N., Hyten, D.L., Arelli, P.R., and Lightfoot, D.A. (2001) 'Forrest' resistance to the soybean cyst nematode is bigenic, saturation mapping of the *Rhg*1 and *Rhg*4 loci. Theor. Appl. Genet. *103*, 710–717.

Njiti, V.N., Gray, L.E., and Lightfoot, D.A. (1997) Rate reducing resistance to *Fusarium solani* f. sp. *phaseoli* underlies field resistance to soybean Sudden Death Syndrome (SDS) Crop Sci. *37*, 132–138.

Njiti, V.N., Doubler, T.W., Suttner, R.J., Gray, L.E., Gibson, P.T., and Lightfoot, D.A. (1998) Resistance to soybean sudden death syndrome and root colonization by *Fusarium solanif*. sp. *glycines* in near-isogeneic lines. Crop Sci. *38*, 472–477.

Njiti, V.N., Johnson, J.E., Torto, T.A., Gray, L.E., and Lightfoot, D.A. (2001) Inoculums' rate influences selection for field resistance to soybean sudden death syndrome in the greenhouse. Crop Sci. *41*, 1726–1731.

Njiti, V.N., Meksem, K., Iqbal, M.J., Johnson, J.E., Kassem, M.A., Zobrist, K.F., Kilo, V.Y., and Lightfoot, D.A. (2002) Common loci underlie field resistance to soybean sudden death syndrome in Forrest, Pyramid, Essex, and Douglas. Theor. Appl. Genet. *104*, 294–300.

Prabhu, R.R., Njiti, V.N., Bell-Johnson, B., Johnson, J.E., Schmidt, M.E., Klein, J.H., and Lightfoot, D.A. (1999) Selecting soybean cultivars for dual resistance to soybean cyst nematode and sudden death syndrome using two DNA markers. Crop Sci. *39*, 982–987.

Ranocha, P., Chabannes, M., Chamayou, S., Danoun, S., Jauneau, A., Boudet, A.M., and Goffiner, D. (2002) Laccase down-regulation causes alterations in phenolic metabolism and cell wall structure in poplar. Plant Physiol. *129*, 145–155.

Ranocha, P., McDougall, G., Hawkins, S., Sterjiades, R., Borderies, G., Stewart, D., Cabanes-Macheteau, M., Boudet, A., and Goffiner, D. (1999) Biochemical characterization, molecular cloning and expression of laccases-a highly divergent gene family in Poplar. Eur. J. Biochem. *259*, 485–495.

Ruben, E., Aziz, J., Afzal, A.J., Njiti, V.N., Triwitayakorn, K., Iqbal, M.J., Yaegashi, S., Arelli, P.R., Town, C.D., Meksem, K., and Lightfoot, D.A. (2006) Genomic analysis of the 'Peking' *rhg1* locus, candidate genes that underlie soybean resistance to the cyst nematode Molec. Genet. Genom. *276*, 320–330.

Searle, I.R., Men, A.E., Laniya, T.S., Buzas, D.M., Iturbe-Ormaetxe, I., Carroll, B.J., and Gresshoff, P.M. (2003) Long-distance signaling in nodulation directed by a CLAVATA1-like receptor kinase. Science *299*, 109–112.

Shopinski, K.L., Iqbal, M.J., Afzal, A.J., Shultz, J.L., Jayaraman, D., and Lightfoot, D.A. (2006) Development of a pooled probe method for locating small gene families in a physical map of soybean using stress related paralogues and a BAC minimum tile path. Plant Methods *2*, 20–28.

Shultz, J.L., Jayaraman, D., Shopinski, K.L., Iqbal, M.J., Kazi, S., Zobrist, K., Bashir, R., Yaegashi, S., Lavu, N., Afzal, A.J., Yesudas, C.R., Kassem, M.A., Wu, C., Zhang, H.B., Town, C.D., Meksem, K., and Lightfoot, D.A. (2006a) The soybean genome database (SoyGD), A browser for display of duplicated, polyploid, regions and sequence tagged sites on the integrated physical and genetic maps of *Glycine max*. Nucleic Acids Res. *34*, D1-D8.

Shultz, J.L., Yesudas, C.R., Yaegashi, S., Afzal, A.J., Kazi, S., and Lightfoot, D.A. (2006b) Three minimum tile paths from bacterial artificial chromosome libraries of the soybean (*Glycine max* cv 'Forrest'), Tools for structural and functional genomics. Plant Methods *2*, 9–18.

Sterjiades, R., Dean, J.F.D., Gamble, G., and Himmelsbach, D.S. (1993) Extracellular laccases and peroxidases from sycamore maple (*Acer pseudopplatanus*) cell suspension cultures Reactions with monolignols and lignin model compounds. Planta *190*, 75–87.

Triwitayakorn, K., Njiti, V.N., Iqbal, M.J., Yaegashi, S., Town, C.D., and Lightfoot, D.A. (2005) Genomic analysis of a region encompassing *QRfs1* and *QRfs2*, genes that underlie soybean resistance to sudden death syndrome. Genome *48*, 125–138.

Wu, C., Sun, S., Nimmakayala, P., Santos FA., Springman, R., Ding, K., Meksem, K., Lightfoot, D.A., and Zhang, H.B. (2004) A BAC and BIBAC-based physical map of the soybean genome. Genome Res. *14*, 319–326.

Zuyun, H., Houping, H., Ruxiu, C., and Yun'e, Z. (1998) Organic solvent enhanced spectrofluorimetric method for determination of laccase activity. Anal. Chim. Acta *374*, 99–103.

Supplementary Material

```
Query   1097  TTTAACATATAATTATACTTGGGAAAAATTTACTAGAAATAGTAATAATAGATTCTCTAA  1156
              ||||||||||||||||||||||||||||||||||||||||||||||||||||||||||||
Sbjct  57768  TTTAACATATAATTATACTTGGGAAAAATTTACTAGAAATAGTAATAATAGATTCTCTAA  57827

Query   1157  CACTTTCTCCTAACATAGTCTATGATTAATTTAAATTTATTGAAAACTATGAAGTTATGA  1216
              ||||||||||||||||||||||||||||||||||||||||||||||||||||||||||||
Sbjct  57828  CACTTTCTCCTAACATAGTCTATGATTAATTTAAATTTATTGAAAACTATGAAGTTATGA  57887

Query   1217  GAGAATTATTATTTTTGTAATTTTTAAGAAATTTCAACTAGAGAAAATATGTTTAAAAGA  1276
              ||||||||||||||||||||||||||||||||||||||||||||||||||||||||||||
Sbjct  57888  GAGAATTATTATTTTTGTAATTTTTAAGAAATTTCAACTAGAGAAAATATGTTTAAAAGA  57947

Query   1277  GCCTGCTGTTAAACTTTCTCAAAATTTATTTTCAACCTCTAACCGCAGACTTCTGAAATA  1336
              ||||||||||||||||||||||||||||||||||||||||||||||||||||||||||||
Sbjct  57948  GCCTGCTGTTAAACTTTCTCAAAATTTATTTTCAACCTCTAACCGCAGACTTCTGAAATA  58007

Query   1337  AGCATTCATGCACTTTATTAACTAGCGGGTGCAACAACTCCTATGGGTTTGGAACCAAGT  1396
              ||||||||||||||||||||||||||||||||||||||||||||||||||||||||||||
Sbjct  58008  AGCATTCATGCACTTTATTAACTAGCGGGTGCAACAACTCCTATGGGTTTGGAACCAAGT  58067

Query   1397  TAAGTTTCCCTTTGGGGTCTGACCTCAACTAAATTAACCTAATCTGCCTAACCTCAAAGG  1456
              ||||||||||||||||||||||||||||||||||||||||||||||||||||||||||||
Sbjct  58068  TAAGTTTCCCTTTGGGGTCTGACCTCAACTAAATTAACCTAATCTGCCTAACCTCAAAGG  58127

Query   1457  ACTTTATCTTTCCCCCACCTCTAATCCACCCTATAAAAGCACCCTCTCCCACTCTTACTT  1516
              ||||||||||||||||||||||||||||||||||||||||||||||||||||||||||||
Sbjct  58128  ACTTTATCTTTCCCCCACCTCTAATCCACCCTATAAAAGCACCCTCTCCCACTCTTACTT  58187

Query   1517  GCATTGCAACCTTAACCTTCAGCATTCACACTAAGGTGTTCCTTGCTCGCCAAAAGATCA  1576
              ||||||||||||||||||||||||||||||||||||||||||||||||||||||||||||
Sbjct  58188  GCATTGCAACCTTAACCTTCAGCATTCACACTAAGGTGTTCCTTGCTCGCCAAAAGATCA  58247

Query   1577  TGGAGCCTGCCAAAACCATTCACAACAATGTCAAATACTCCCCCATCTTCTTAGCCATCT  1636
              ||||||||||||||||||||||||||||||||||||||||||||||||||||||||||||
Sbjct  58248  TGGAGCCTGCCAAAACCATTCACAACAATGTCAAATACTCCCCCATCTTCTTAGCCATCT  58307

Query   1637  TTGTTCTGATCTTAGCTTCAGCATTGTCTTCAGCAAATGCCAAGATTCACGAGCACGAGT  1696
              ||||||||||||||||||||||||||||||||||||||||||| ||||||||||||||||
Sbjct  58308  TTGTTCTGATCTTAGCTTCAGCATTGTCTTCAGCAAATGCCAAAATTCACGAGCACGAGT  58367

Query   1697  TAGTAGTACAATCTCGCACTCTGCAAGCTCAGAAGGGCTCAGGACGGATGAGACGCAAGT  1756
              | || ||||| |||| |||||| |   |||  ||   ||| ||||  ||   |  |  ||
Sbjct  58368  TTGTTGTACATTCTCTCACTCT-CTTTCTCTTAATTTCTCTGGACTTATTTTATTCTTGT  58426

Query   1757  CGATAAGAACTCAGATCCGCGAGATAATAATGACCTAGCTGAAGATAGTAATGGAATAGG  1816
                 |   |||||   ||||  |||||||||| |||||||||  ||| |||||||||||||
Sbjct  58427  TTTTTTTAACTCTTTTCCGTTAGATAATAATTACCTAGCTGTTGATTGTAATGGAATAGG  58486

Query   1817  AAGAAGCAACTCCAGTGAAGAGGCTGTG-AAAACCCACAACAGCATCACCGTGAATGGAC  1875
                |||||||||||||||||||||||||| |||||||||||||||||||||||||||||||
Sbjct  58487  TTGAAGCAACTCCAGTGAAGAGGCTGTGCAAAACCCACAACAGCATCACCGTGAATGGAC  58546
```

Supplemental Fig. 1. Alignment of the DNA sequence of Williams 82 and A3244 in the promoter region, first exon and first intron of the laccase at the *Rfs2/ rhg1* locus.

Chapter 3

The Role of Green Fluorescent Protein (GFP) in Transgenic Plants to Reduce Gene Silencing Phenomena

Hany A. El-Shemy, Mutasim M. Khalafalla, and Masao Ishimoto

Abstract

The green fluorescent protein (GFP) of jellyfish (*Aequorea victoria*) has significant advantages over other reporter genes, because expression can be detected in living cells without any substrates. Recently, epigenetic phenomena are important to consider in plant biotechnology experiments for elucidate unknown mechanism. Therefore, soybean immature cotyledons were generated embryogenesis cells and engineered with two different gene constructs (pHV and pHVS) using gene gun method. Both constructs contain a gene conferring resistance to hygromycin (*hpt*) as a selective marker and a modified glycinin (11S globulin) gene (*V3–1*) as a target. However, *sGFP(S65T)* as a reporter gene was used only in pHVS as a reporter gene for study the relation between using *sGFP(S65T)* and gene silencing phenomena. Fluorescence microscopic was used for screening after the selection of hygromycin, identified clearly the expression of *sGFP(S65T)* in the transformed soybean embryos bombarded with the pHVS construct. Protein analysis was used to detect gene expression overall seeds using SDS-PAGE. Percentage of gene down regulation was highly in pHV construct compared with pHVS. Thus, *sGFP(S65T)* as a reporter gene in vector system may be play useful role for transgenic evaluation and avoid gene silencing in plants for the benefit of plant transformation system.

Introduction

The advent of the technology to transform plants with DNA sequences such as the generation of transgenic plants, allowed an entirely new direction for the exploration of genetics.

GFP has grown in popularity as a reporter gene in molecular biology research. In plant transformation systems with the problem of low efficiency, the nondestructive analysis of putative transformed cells and tissues using a reporter gene such as the GFP (jellyfish green fluorescence protein) gene could significantly help to optimize transformation protocols and generate transgenic plants (Stewart, 2001; El-Shemy *et al.*, 2004, 2006, 2007).

With the successful development of procedures for transformation, transgenic plants and their progeny have been studied for continued stable expression of the foreign gene through several generations. These studies have revealed that in some cases the transgene expression was lost in a variable proportion of the progeny. This phenomenon, called "gene silencing", has been studied most extensively in dicot plants, such as tobacco, petunia, tomato, and buckweed (Matzke and Matzke, 1995; Meyer, 1995). Several mechanisms have been proposed to explain the phenomena of gene silencing (Matzke *et al.*, 2001; Meyer, 1995; Stam *et al.*, 1997). Two types of silencing occur in plants and in other eukaryotes: transcriptional gene silencing (TGS) and posttranscriptional gene silencing (PTGS) (Carthew,2001; Waterhouse *et al.*, 2001a,b). Silencing at the transcriptional level is thought to occur primarily by methylation of promoter sequences, thereby interfering with assembly of the transcription factors and/or by attracting chromatin re-modeling proteins to these sites (Meyer, 2000; Wang and Waterhouse, 2002).

Gene silencing and its implications for transgene expression is an area of intense research at this time, and the reader is directed to recent reviews on this large subject (Meyer, 2000; Vance and Vaucheret, 2001; Wang and Waterhouse, 2002). For that, the aim of this study is to identify the role of *sGFP(S65T)* to reduce or avoid the gene silencing phenomenon in transgenic plants.

Results

Embryogenic cells induced from a soybean cultivar, Jack, were transformed by microparticle bombardment with the pHVS, which contains a modified globulin gene, a selectable marker gene *hpt* and a reporter gene *sGFP(S65T)*. To optimize the conditions for particle delivery into the embryogenic tissues, transient expression of the *sGFP(S65T)* gene was detected one day after bombardment with a fluorescent microscope. According to the intensity and number of foci expressing

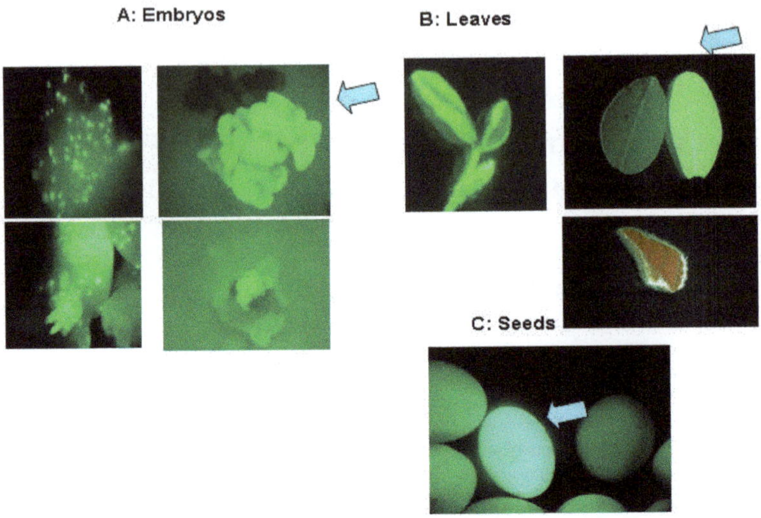

Fig. 1. Visualization of *sGFP(S65T)* expression in transformed soybean plantlets. Embryos (**A**), leaves (**B**) and seeds (**C**).

GFP, we decided on a pressure of 1350 psi and distance of 6 cm for the delivery. Expression of *sGFP(S65T)* in soybean was also monitored during the selection with hygromycin and development of plants (Fig. 1).

In this experiment, the conditions for delivery of particles coated with plasmids according to the transient expression was successfully optimized of the *sGFP(S65T)* gene (Fig. 2).

Transformation was achieved by coating each plasmid, either pHV or pHVS, onto the particles and bombarding embryogenic tissues. Hygromycin-resistant cells were selected then matured in FNL0S3S3 liquid medium and germinated on MS0 media. All regenerated plants obtained from the two constructs were confirmed the presence of *hpt* gene by PCR analysis. Out of a total of 122 regenerated plants obtained from the introduction of pHVS, 82 plants produced an expected band with a 0.5 kbp of PCR product within the *hpt* gene (Table 1). On the other hand, 29 of 98 regenerated plants obtained from the introduction of pHV yielded the 0.5 kbp *hpt* fragment in PCR analysis (Table 1). PCR analysis for *sGFP(S65T)* genes was conducted in the soybean plants that yielded a *hpt* band to confirm the presence of all transgene cassettes (Fig. 2). The expression of sGFP(S65T) was detected in about 52 % of the *hpt*-positive soybeans engineered using pHVS (Fig. 2). Southern blot analysis was performed to confirm the integration, and to estimate the copy numbers of transgenes. Total genomic DNA, which was isolated from transgenic plants to as certain the presence of all transgene cassettes by PCR (Fig. 2), was digested with *SacI*, and hybridized with one of the two probes for *hpt* and *sGFP(S65T)* (Fig. 3). All the transformants analyzed here yielded one to seven bands hybridized with the *hpt* probe in addition to a common band at around 6.6 kbp (Fig. 3A). On the other hand, DNA isolated from untransformed plants hybridized with the *sGFP(S65T)* probe, even though GFP is not derived from plants. This maybe caused by unspecific hybridization of the probe with soybean genomic DNA (Fig. 3B). The individual plants exhibited different banding patterns, confirming that they resulted from different events.

Table 1 Gene silencing efficiency in soybean by using two vector systems.

Construct	pHVS	pHV
Regenerated plants	122	98
hpt-positive by PCR	82	29
Transgenic plants[a]	49 (29[b]/20[c])	21
Absence of glycinin	24 (10[b]/14[c])	15
Gene Silencing %	58.5 (34.4[b]/70[c])	71.4

[a]Transformats giving *hpt* and *sGFP* bands by PCR analysis
[b]Transgenic soybean expressing s*GFP*
[c]Transgenic soybean not expressing s*GFP*

The accumulation of glycinin was confirmed by SDS-PAGE and Tricine SDS-PAGE analysis of the globulin fraction extracted from transgenic seeds (Fig. 4). The modified glycinin *V3–1* could not be distinguished from endogenous glycinin subunits by the SDS-PAGE, because the modified glycinin contains only six additional amino acids in the basic subunit. The results of protein analysis were confirmed by immunoblotting (Fig. 5).

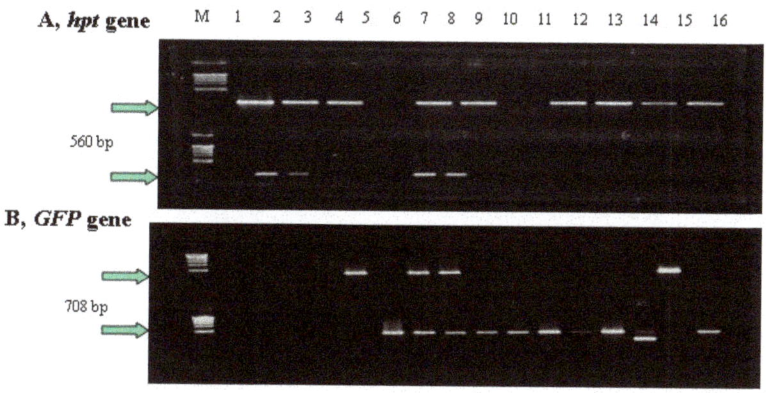

Fig. 2. Detection of foreign genes in transgenic soybean plants by PCR: *hpt* (**A**), and *sgfp* (**B**) genes. lane 1 (untransformed control), 2–15 (transformed soybeans with pHVS and pHV), and 16 (plasmid with pHVS). Arrowheads: expected 560 bp and 708 bp fragments of *hpt* and *sgfp* genes.

Discussion

The potential of GFP appears to be greater than its current applications as researchers seeking to characterize GFP under novel conditions reveal new uses for the protein (Stewart, 2001; Richards *et al.*, 2003; El-Shemy *et al.*, 2004, 2006). This increased potential is especially pertinent to plants, since it is often desirable to quantify gene expression on the plant and in the field.

GFP can partially replace antibiotic selection and be of great use when the organogenesis or conversion of transformation procedures is inefficient under antibiotic or herbicide selection (Stewart, 2001). It could be helpful in isolating events during the early stages of transformation experiments as described here (Fig. 1). An example of this was the transformation of sugarcane with *sGFP(S65T)* (Elliot *et al.*, 1999).

Silencing was first observed concerning transgene expression in plants about 10 years ago, with a report that transformation of petunia with extra copies of the chalcone synthase gene could result in a block in expression of both the transgene and the corresponding endogenous gene (Napoli *et al.*, 1990; Lessard *et al.*, 2002). This phenomenon was termed cosuppression (Napoli *et al.*, 1990). Suppression of endogenous glycinin

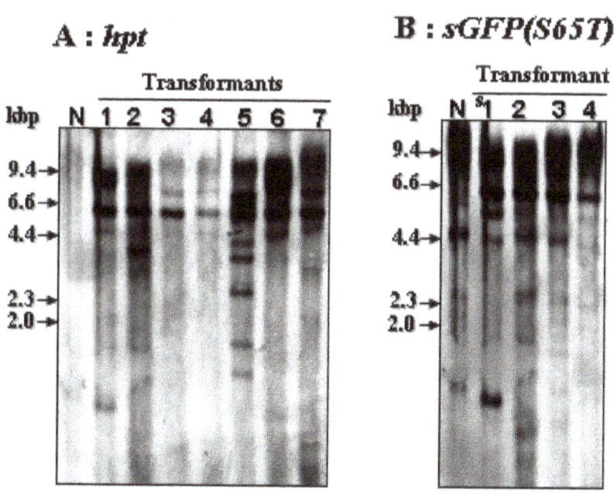

Fig. 3. Southern blot analysis of soybeans transformed with pHVS and pHV. Total DNA was digested with *SacI* and loaded in lane N (untransformed control), 1–4 (transformed soybeans with pHVS), and 5–7 (transformed soybeans with pHV). Separated DNA was transferred to a nylon membrane and hybridized with each probe containing the coding region of *hpt* (**A**), and *sGFP(S65T)* (**B**).

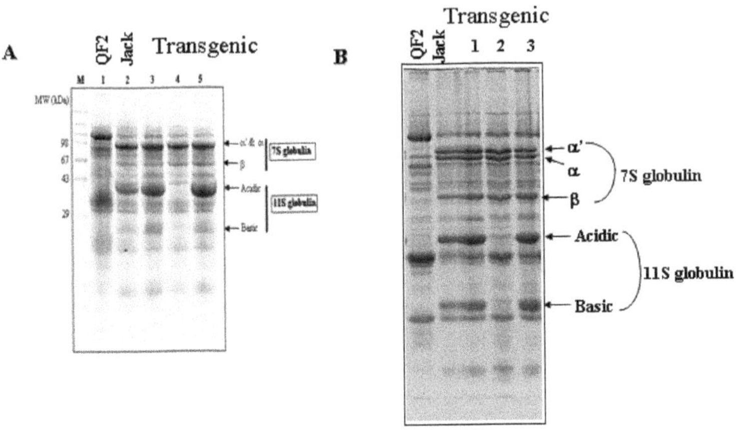

Fig. 4. Tricine-SDS-PAGE (**A**) and SDS-PAGE (**B**) analysis of components of seed storage proteins in transformed soybean. Globulin fractions were isolated from QF2 (mutant line lacking 7S and 11S globulin), Jack (untransformed control) and transformants, 25 µg of each fraction was fractionated by SDS-PAGE and then stained with CBB.

in the transformed soybean was frequently observed on the introduction of the modified glycinin *V3–1* gene (Table 1).

The effects of transgenic copy number on the level of gene expression are known to be complex. Though it was anticipated that the increase of transgene copy number would increase the expression level (Dai *et al.*, 2001; El-Shemy *et al.*, 2004, 2006, 2007), it is now known that gene co-suppression phenomena frequently occur in transgenic plants with repeated transgenes or an unusual structure such as inverted repeats (Vaucheret *et al.*, 1998). Such events are likely to be powerful inducers of co-suppression and methylation (Luff *et al.*, 1999).

The glycinin subunit polypeptides in some transformants were intensely stained with CBB compared to non-transformants (Fig. 4). This may be due to the accumulation of the modified glycinin *V3–1* in transgenic seeds. On the other hand, some transgenic soybeans lack all subunits of glycinin, suggesting the transgene may cause the suppression of endogenous glycinin genes by the effect of gene silencing (Table 1, Fig. 4 and Fig. 5). The ratio of gene silencing was lower in transformants engineered with the pHVS construct and selected based on GFP expression than transgenic soybean engineered with the pHV construct (Table 1).

It is suggested that the transgene causes the effects of gene silencing. There is some predictability to silencing. Counterintuitively, an increased copy number of a transgene can correlate with an increased risk of silencing (Lessard *et al.*, 2002). However, determining whether silencing will occur in particular transgenic plants is still largely an empirical problem, requiring testing of transgene expression in the individual plants. Two general approaches can be used to avoid problems with silencing (Lessard *et al.*, 2002). First, the use of gene delivery methods, such as *Agrobacterium*-mediated transformation, that result in integration of relatively few copies of a transgene into the genome,can minimize problems with silencing (Dai *et al.*, 2001). Second,the use of constructs in which matrix attachment regions

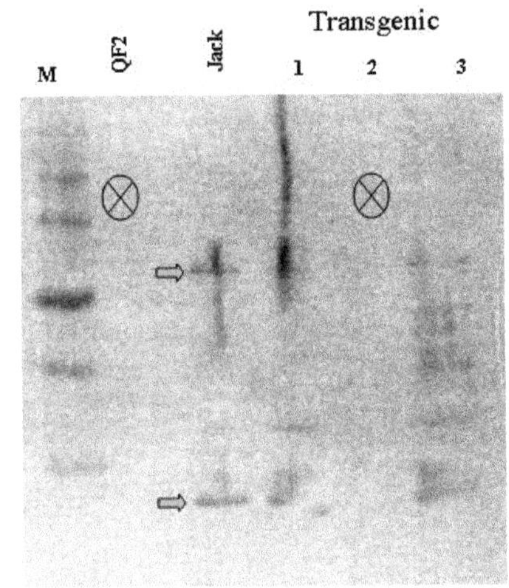

Fig. 5. Detection of storage proteins (11S globulin) in soybean transgenic seed proteins (Immune-blotting). QF2 (mutant line lacking 7S and 11S globulin), Jack (untransformed control) and transformants. lane 1 (transformed soybeans with pHVS), lane 2(transformed soybeans with pHV) and lane 3 (transformed soybeans with pHVS).
Arrowheads expected fragments of 11S globulin (Jack, untransformed control, lane 1, 3 transformed soybeans with pHVS).
QF2 and Lane 2 transformed soybeans with pHV showed silencing of internal 11S globulin.

flank the transgene may also minimize silencing (Spiker and Thompson,1996). The results expected that the transgenic soybeans accumulating the modified glycinin *V3–1* to have a higher level of methionine than nontransformants (Fig. 4). Up to date, gene silencing was seen as a problem for plant genetic transformation, as it prevented reliable expression of a desired phenotype within transgenic plants (Taylor and Fauquet, 2002). However, with increasing knowledge of the mechanisms underlying this phenomena, and realization that it can be utilized to down-regulate native genes within the plants, and it will become a powerful tool in future transgenic applications (Vance and Vaucheret, 2001; Lessard *et al*,, 2002).

The silencing effect appeared stochastic in nature, with patterns and variegation appearing in the tissues of the transformed plants, much like previous observations of somaclonal variation and paramutation (Grant-Downton and Dickinson, 2005). This 'co-suppression' of an endogenous gene by extra copies of the same gene was also accompanied by observations that insertion of entirely foreign coding gene sequences did not necessarily lead to their expression (Grant-Downton and Dickinson, 2005). Other experimentation with transgenic technology led to the discovery that this gene silencing effect on endogenous genes by inserted DNA was more consistent when the inserted gene copy was in reverse or 'antisense' orientation (Van der Kroll *et al.*, 1988). Perhaps more remarkable was the realization that a whole copy of the gene was not even required to elicit this effect (Grant-Downton and Dickinson, 2005). Clearly, adding new fragments of DNA to the genome not only resulted in the silencing of their own expression, but also the specific silencing of endogenous genomic sequences to which they were homologous (Grant-Downton and Dickinson, 2005). An intact copy with no nucleotide mutation was being 'silenced' by the presence of a foreign sequence at another site in the same genome (Grant-Downton and Dickinson, 2005).

Biomolecular techniques allow for the insertion of the gene for GFP right before the stop codon (Chalfie *et al.*, 1994). The cell would keep making hemoglobin, but before reaching the stop codon it would make the GFP (Chalfie *et al.*, 2004; Fig. 6).

There were three reasons Prasher thought that GFP could potentially become a significant tracer molecule. Firstly, if enough protein with attached GFP were made, it should be easy to detect and to trace it as it moved through the cell, because irradiating the cell with ultra violet light would cause the GFP attached to the protein to fluoresce. Secondly, Shimomura had shown in 1974 that GFP was a fairly small protein. This was important because a small protein attached to the protein of interest was less likely to hinder its proper function. Its small size would also allow it to follow the fused protein, especially in organelles like neurons, whereas the diffusion of large proteins would be difficult. Thirdly, it had been shown that once GFP was made in the jellyfish, it was fluorescent. Most other bioluminescent molecules require the addition of other substances before they glow. For example, aequorin will glow only if calcium ions and coelenterazine have been added, and firefly luciferase requires ATP, magnesium, and luciferin before it luminesces. This would make GFP a much more versatile tracer than either aequorin or firefly luciferase, which were being used as tracers. Besides attaching GFP to a protein and making it a fluorescent tag, Prasher also thought that GFP could potentially be a very useful reporter molecule. For activating protein production, DNA promoters are used; these are sequences of DNA next to genes that contain the information about where and when the gene should be read and make the protein. If GFP is linked to

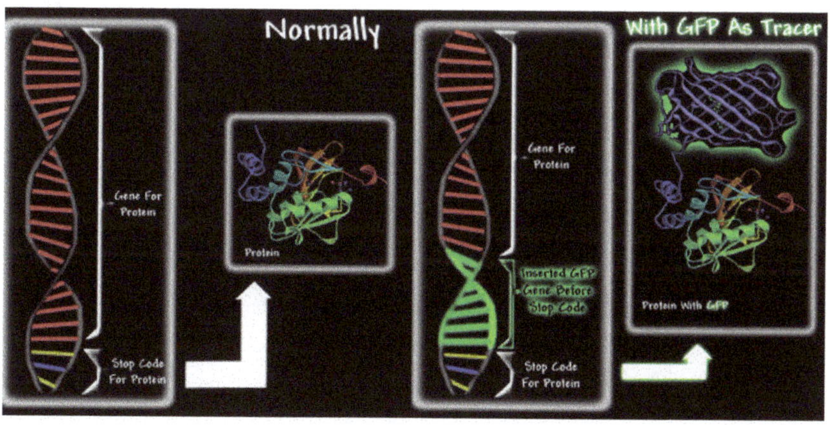

Fig. 6 Insertion of GFP gene into DNA. Douglas Prasher GFP School, (*Science* 263, 802–805, 1994). http://www.conncoll.edu/ccacad/zimmer/GFP-ww/GFP-1.htm

a specific promoter then it will be expressed in place of the protein, showing where and when the gene of interest is switched on (Douglas Prasher, from Zimmer, Marc website, Fig. 6).

Therefore, *sGFP(S65T)* can be effectively used to select the transformants expressing all the gene cassettes (El-Shemy *et al.*2004, 2006, 2007). The transgenic plants expressing *sGFP(S65T)* grew and reproduced normally, and the GFP expression was inherited without any abnormalities. In conclusion, using *sGFP(S65T)* as a reporter gene reduced or avoided the gene silencing in transgenic soybean.

Materials and methods

Construction of chimeric genes

Two plasmid vectors, pHVS and pHV, were constructed based on the pUC19 vector (Fig. 7 A and B). pHV contains the hygromycin phosphotransferase coding region, *hpt* (1.0kb), under regulatory control of the cauliflower mosaic virus (CaMV) 35S promoter, *35Spro*, and the modified proglycinin (A1aB1b) cDNA, *V3–1* (1.4kb), with a synthetic DNA encoding four continuous methionines. pHVS contains additionally a modified jellyfish green fluorescent protein coding region, *sGFP(S65T)* (0.8kb), under regulatory control of *35Spro* in the flanking region of the *V3–1* gene (El-Shemy *et al*,, 2004, 2006, 2007).

Initiation and proliferation of embryogenic cultures

Transformation and regeneration systems for soybean were optimized according to methods described elsewhere (Finer and Nagasawa, 1988; Hadi *et al.*, 1996; Sato *et al.*, 1993; El-Shemy *et al.*, 2004, 2006, 2007). Soybean plants, cv. Jack, were grown in soil in a glass-house controlled at 25°C under natural light. Developing green pods were obtained when the immature cotyledons were about 4–5 mm long. After sterilization of the pod surface with 70% ethanol followed by 3 rinses with sterile water, the immature cotyledons were extracted, the end removed together with the embryonic axis, and the sample placed flat side up on MSD40 medium consisting of MS salts (Murashige and Skoog, 1962) and B5 vitamins (Gamborg *et al.*, 1968) supplemented with 3% sucrose, 40 mg/l 2,4-D, and 0.2% Gelrite (Wako Pure Chemical Industries, Osaka, Japan) (pH7.0). Embryogenic tissues were initiated at 25°C under cool white fluorescent light (23/1 light regime, 5–10 μmol/m^2/s) for 3 to 4 weeks. Soybean embryogenic tissues were suspended and maintained in FN Lite liquid medium consisting of FN Lite macro salts, MS micro salts and B5 vitamins supplemented with 1 g/l asparagine, 5 mg/l 2,4-D, and 1% sucrose (pH 5.8) (Finer and Nagasawa, 1988; El-Shemy *et al.*, 2004), and maintained by subculturing every week into 25 ml of fresh FN Lite liquid medium in a 100 ml flask (El-Shemy *et al.*, 2004, 2006).

Transformation by particle bombardment

Approximately 1 g of embryogenic suspension tissue was transferred to the center of MSD20 medium consisting of MS salts and B5 vitamins supplemented with 3% sucrose, 1 g/l asparagine 20 mg/l 2,4-D, and 0.2% Gelrite (pH5.8) in a 9 cm petri dish. Bombardments were performed using a

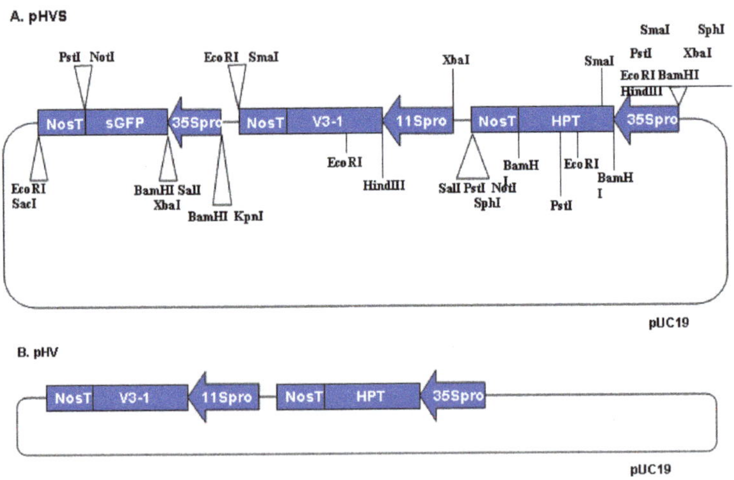

Fig. 7. Structures of plasmid constructs for soybean transformation. **A**, pHVS contains a modified glycinin gene, *V3–1*, between *hpt* as a selectable gene and *sGFP(S65T)* as a reporter gene. **B**, pHV contains *V3–1* flanked by *hpt* as a selectable gene. Restriction sites are indicated.

Biolistic PDS-1000/He Particle Delivery System (Bio-Rad, Richmond, CA, USA) according to the instruction manual. Each sample of embryogenic tissue was bombarded twice with a pressure of 1350 *psi*.

Bombarded tissues were resuspended in the FN Lite medium. One week after bombardment the embryogenic tissues were transferred to fresh FN Lite medium containing 15 mg/l hygromycin B (Roche Diagnostics, Mannheim, Germany). The bombarded tissues were transferred to fresh antibiotic-containing FN Lite medium weekly for an additional 3 weeks. Then, white lumps of tissue that contained bright green lobes of embryogenic tissue were selected and transferred to fresh FN Lite medium containing 30 mg/l hygromycin B. The hygromycin tolerant tissues were selected and resuspended in fresh antibiotic-containing FN Lite medium weekly for 3 additional weeks (El-Shemy *et al.*, 2004, 2006).

Hygromycin tolerant embryos were subcultured in FNL0S3S3 liquid medium, which contained FN Lite macro salts, MS micro salts and B5 vitamins supplemented with 1 g/l asparagine, 3% sucrose, and 3% sorbitol (pH 5.8). Three weeks after the suspension, excess liquid of the developing embryos was withdrawn with sterile filter paper, and the embryos were placed in dry petri dishes for 3 to 5 days. After the desiccation treatment, the embryos were placed on MS0 medium containing MS salts, B5 vitamins, 3% sucrose, and 0.2% Gelrite (pH5.8). The germinating plantlets were transferred to ½ B5 medium. After root and shoot elongation, plantlets were transferred to pots containing soil, and maintained under high humidity. Plantlets were gradually adapted to ambient humidity and placed in the glass-house.

GFP detection

The presence of sGFP(S65T) was detected by blue light excitation (Chiu *et al.*, 1996). Embryos, cotyledons, leaves of regenerated plants and seeds were observed using a microscope (Leica Microsystems, Wetzlar, Germany) with a filter set providing 455–490 nm excitation and emission above 515 nm.

PCR and Southern blot analysis

Total DNA was isolated from soybean leaves by the method of (Draper and Scott, 1988), and used to investigate the presence of the transgenes. PCR analysis was conducted to screen transformed plants in a 20 µl reaction mixture containing 10 ng of genomic DNA, 200 µM of each dNTP, 0.2 µm of each primer, and 2.5 units of Ampli-taq Gold polymerase (Applied Biosystems, Foster City, CA USA) in the corresponding buffer. Reaction were hot-started (9 min at 94°C) and subjected to 30 cycles as follows: 30 sec at 94°C; 1 min at 55°C; and 1 min at 72°C. The last extension phase was prolonged to 7 min at 72°C. The primer set for *hpt* was designed for amplification of a 560 bp fragment; sequences are 5′-ATCCTTCGCAAGACCCTTCCT-3′ (35S promoter) and 5′-GGTGTCGTCCATCACAGTTTG-3′ (*hpt*). The primer set for *sGFP(S65T)* was designed for amplification of a 708 bp fragment; sequences are 5′- AAGGTACCGGATCCCCCCTCAGAA-3′ and 5′-AAGAGCTCCGATCTAGTAACATAGATGACACC-3′.

Southern blot analysis was conducted to confirm the stable integration of transgenes into soybean. Total DNA (10 µg) was digested with the restriction enzyme, *SacI*, and digested DNA was separated by electrophoresis in a 1% agarose gel and transferred onto a hybond N+ membrane (Amersham Biosciences, Buckinghamshire, England). Labeling and detection were conducted following the protocol of ECL direct nucleic acid labeling and detection (Amersham Biosciences). DNA fragments of the *sGFP(S65T)* gene and *hpt* gene were amplified from the plasmids with the same primer sets used for the PCR analysis, and served as hybridization probes on Southern blot membranes.

Analysis of seed proteins in transgenic soybean

The subunit composition of seeds from individual transgenic soybean was analyzed by SDS-PAGE (Laemmli, 1970). A total globulin fraction was extracted from soybean seed meal by grinding with 50 mM Tris-HCl, pH 8.0. The supernatant was obtained after centrifugation at 15, 000 *g* for 10 min. The protein concentration was determined with a BCA protein assay (Pierce, Rockford, IL, USA) kit using bovine serum albumin as a standard. The protein solution was mixed with the same amount of a twice-concentrated SDS sample buffer containing 2-mercaptoethnol. The proteins (25 µg) from each sample were separated on a gels (Tricine-SDS-PAGE and SDS-PAGE) containing 12 % (w/v) acrylamide and 0.2 % (w/v) bis-acrylamide, and were stained with Bio-Safe CBB G-250 stain (Bio-Rad).

Acknowledgements

Hany A. El-Shemy (Cairo University, Faculty of Agriculture, Biochemistry Department, Giza, Egypt) was supported by a postdoctoral fellowship from the Japanese Science and Technology Corporation. Mutasim M. Khalafalla (Commission for Biotechnology and Genetic Engineering, Sudan) was supported by a postdoctoral fellowship from the Japan Society for the Promotion of Science. The authors thank Yumi Nakamoto, Yumi Naganuma and Midori Yokoyama for technical assistance.

References

Carthew, R. W. (2001) Gene silencing by double-stranded RNA. Curr. Opin. Cell Biol. *13*, 244–248.

Chalfie, M., Tu, Y., Euskirchen, G., Ward, W.W. & Prasher, D.C. (2004) Green fluorescent protein as a marker for gene expression. Science, *263*, 802–805.

Chiu, W-L., Niwa, Y., Zeng, W., Hirano, T., Kobayashi, H., and Sheen, J. (1996) Engineered GFP as a vital reporter in plants. Curr. Biol. *6*, 325–330.

Dai, S., Zheng, P., Marmey, P., Zhang, S., Tian, W., Chen, S., Beachy, R.N., and Fauquet, C. (2001) Comparative analysis of transgenic rice plants obtained by *Agrobacterium*-mediated transformation and particle bombardment. Mol. Breed. *7*, 25–33.

Draper, J., and Scott, R. (1988) The isolation of plant nucleic acids. In: Plant genetic transformation and gene expression, Draper, J., Scott, R., Armitage, P., Walden, R., eds. (London: Blackwell Scientific Publications), pp. 212–214.

Elliot, A., Campbell, J., Brettell, I., and Grof, P. (1998) *Agrobacterium*-mediated transformation of sugarcane using GFP as a screenable marker. Aust. J. Plant Physiol. *25*, 739–743.

Elliot, A., Campbell, J., Dugdale, B., Brettell, R., and Grof, C. (1999) Green-fluorescent Protein facilitates rapid *in vivo* detection of genetically transformed plant cells. Plant Cell Rep. *18*, 707–714.

El-Shemy, H.A., Khalafalla, M.M., Nishizawa, K., Utsumi, S., and Ishimoto, M. (2004) Isolation of soybean plants with stable transgene expression by visual selection based on green fluorescent protein. Mol. Breed. *14*, 227–238.

El-Shemy, H. A, Khalafalla, M.M., Fujita, K., and Ishimoto, M. (2006) Molecular control of gene co-suppression of transgenic soybean. J. Biochem. Mol. Biol. *39*, 61–67.

El-Shemy, H. A, Khalafalla, M.M., Fujita, K., and Ishimoto, M. (2007) Improvement of protein quality in transgenic soybean plants. Biol. Plant. *51*, 277–284.

Finer, J., and Nagasawa, A. (1988) Development of an embryogenic suspension culture of soybean [*Glycine max* (L.) Merrill]. Plant Cell Tiss. Org. Cult. *15*, 125–136.

Gamborg, O., Miller, R., and Ojima, K. (1968) Nutrient requirements of suspension cultures of soybean root cell. Exp. Cell Res. *50*, 151–158.

Grant-Downton, R.T., and Dickinson, H.G. (2005) Epigenetics and its implications for plant biology. The epigenetic network in plants. Ann. Bot. *96*, 1143–1164.

Hadi, M.Z., McMullen, M.D., and Finer J.J. (1996) Transformation of 12 different plasmids into soybean via particle bombardment. Plant Cell Rep. *15*, 500–505.

Laemmli, U.K. (1970) Cleavage of structural proteins during the assembly of the head bacteriophage T4. Nature *277*, 680–685.

Lessard, P.A., Kulaveerasingam, H., York, G.M., Strong, A., and Sinskey, A.J. (2002) Manipulating gene expression for the metabolic engineering of plants. Metab. Eng. *4*, 67–79.

Luff, B., Pawlowski, L., and Bende, J. (1999) An inverted repeat triggers cytosine methylation of identical sequences in *Arabidopsis*. Mol. Cell. *3*, 505–511.

Matzke, A., and Matzke, M. (1995) Trans-inactivation of homologous sequences in *Nicotiana tabacum*. In: Gene silencing in higher plants and related phenomena in other eukaryotes, Meyer, P. ed. (Berlin: Springer), pp1–14.

Matzke, M., Matzke, A.J.M., and Kooter, J.M. (2001) RNA: guiding gene silencing. Science *293*, 1080–1083.

Meyer, P. (1995) Understanding and controlling transgene expression. Trends Biotechnol. *13*, 332–337.

Meyer, P. (2000) Transcriptional transgene silencing and chromatin components. Plant Mol. Biol. *43*, 221–234.

Murashige, T., and Skoog, F. (1962) A revised medium for rapid growth and bioassays with tobacco tissue culture. Physiol. Plant *15*, 473–497.

Napoli, C.,Lemieux, C.,and Jorgensen, R.A. (1990) Introduction of a chimeric chalcone synthase gene in petunia results in reversible cosuppression of homologous genes *in trans*. Plant Cell *2*, 279–289.

Richards, H.A., Halfhill, M.D., Millwood, R.J., and Stewart, C.N. (2003) Quantitative GFP fluorescence as an indicator of recombinant protein synthesis in transgenic plants. Plant Cell Rep. *22*, 117–121.

Sato, S., Newell, C., Kolacz, K., Tredo, L., Finer, J., and Hinchee, M. (1993) Stable transformation via particle bombardment in two different soybean regeneration systems. Plant Cell Rep. *12*, 408–413.

Spiker, S., and Thompson,W. F. (1996) Nuclear matrix attachment regions and transgene expression in plants. Plant Physiol. *110*, 15–21.

Stam, M, Mol, J., and Kooter, J. (1997) The silence of genes in transgenic plants. Ann. Bot. *79*, 3–12.

Stewart, J.C.N. (2001) The utility of green fluorescent protein in transgenic plants. Plant Cell Rep. *20*, 376–382.

Taylor, N.G., and Fauquet, C.M. (2002) Microparticle bombardment as a tool in plant science and agricultural biotechnology. DNA Cell Biol. *21*, 963–977.

Van der Krol, A., Lenting, P., Veenstra, J,. van der Meer, I., Gerats, A., Mol, J., and Stuitje, A. 1988. An anti-sense chalcone synthase gene in transgenic plants inhibits flower pigmentation. Nature *333*, 866–869

Vance, V., and Vaucheret, H. (2001) RNA silencing in plants – defense and counter defense. Science *292*, 2277–2280.

Vaucheret, H., Beclin, C., Elmayan, T., Feuerbach, F., Godon, C., Morel, J-B., Mourrain, P., Palauqui, J.-C., and Vernhettes, S. (1998) Transgene-induced gene silencing in plants. Plant J. *16*, 651–659.

Wang, M-B., and Waterhouse, P. (2002) Application of gene silencing in plants. Curr. Opin. Plant Biol. *5*, 146–150.

Waterhouse, P.M.,Wang,M., and Finnegan, E.J. (2001a) Role of short RNAs in gene silencing. Trends Plant Sci. *6*, 297–301.

Waterhouse,P. M., Wang,M. B., and Lough,T. (2001b) Gene silencing as an adaptive defence against viruses. Nature *411*, 834–842.

Zimmer, M. (2007) Green Fluorescent Protein. http://www.conncoll.edu/ccacad/zimmer/GFP-ww/GFP-1.htm (accessed 4 December 2007).

Chapter 4

Antisense Phenotypes Reveal a Functional Expression of *OsARF1*, an Auxin Response Factor, in Transgenic Rice

Kotb A. Attia, Amr F. Abdelkhalik, Megahed H. Ammar, Chun Wei, Jinshui Yang, David A. Lightfoot, Wagih M. El-Sayed and Hany A. El-Shemy

Abstract

OsARF1 is the first full-length member of auxin response factor (*ARF*) gene family to be cloned from monocot plant. Using quantitative RT-PCR this study found that, the transcript abundance of *OsARF1* was significantly higher in embryonic tissues than in vegetative tissues. To investigate the effect of *OsARF1* on the phenotype of rice, a cDNA fragment of *OsARF1* was inserted in inverse orientation to the 35S promoter in vector pBin438 to produce an antisense (AS) construction. The *AS-OsARF1* construct was transferred into rice (*Oryza sativa* L. *japonica*) calli via *Agrobacterium tumefaciens*-mediated transformation. Molecular analysis of transgenic plants showed that the functional expression of *OsARF1* was inhibited at mRNA level efficiently. The *AS-OsARF1* plants showed extremely low growth, poor vigor, short curled leaves and tillered but were sterile. Therefore, the *OsARF1* was shown to be essential for growth in vegetative organs and seed development.

Abbreviations: ARF, auxin response factor; RT-PCR, reverse transcriptional polymerase chain reaction; AS-DNA/RNA, antisense DNA/RNA; RACE, rapid amplification of cDNA end.

Introduction

Auxin is one of the two most important plant hormones, and regulates various growth and developmental processes by controlling the expression of auxin-response genes (Ulmasov *et al.*, 1995). Auxin responsiveness is conferred to several genes by conserved promoter elements, termed 'auxin-responsive elements' (AuxRE) AuxRE promoter elements are bound by a new class of plant-specific transcription factors, named auxin response factors (ARFs) (Ulmasov *et al.*, 1997a). Because of the extremely low expression of genes encoding ARFs, none were isolated until 1997. The first auxin response factor (*ARF1*) was isolated from *Arabidopsis* (Ulmasov, *et al.*, 1997a). It was found that all members of the *ARF* gene family have an amino-terminal DNA-binding domain and most contain a C-terminal region with two conserved domains, which are involved in homo-and hetero-dimerization (Ulmasov *et al.*, 1999). In *Arabidopsis*, it has been reported that the ARF proteins are encoded by a gene family with 23 members and some of them have been shown to repress or to activate expression of reporter genes with an AuxRE promoter element (Ulmasov *et al.*, 1999a; Remington *et al.*, 2004; Okushima *et al.*, 2005). Recently, the first full-length *ARF* gene of a monocot plant was cloned from rice (Frank *et al.*, 2002). Several rice ARF family transcriptional regulators were also identified homologous to *Arabidopsis ARF1* (Waller *et al.*, 2002). From genome sequences many *ARF* genes in *Arabidopsis* and in rice have been isolated (Akila *et al.*, 2004; Wang *et al.*, 2007). Many of the *ARF* genes mediate up-regulation by auxin but some cause down-regulation by auxin (Ulmasov *et al.*, 1999). Different amino acid biases in mediate regions of *ARF* genes family were shown to underlie the ability to cause up or down regulation to downstream genes, by fusion protein experiments (Ozga *et al.*, 2002). The protein product of a rice auxin response factor *OsARF1* was shown to be localized in nucleus with the OsARF1: GFP fusion protein (Waller *et al.*, 2002).

The data on *ARF* genes family was accumulating rapidly in last few years, but there are still few reports on their effects on phenotype. Most (13/18) *ARF* genes when knocked out in *A. thaliana* plants showed no obvious phenotypic change except the previously identified *arf2/hss*, *arf3/ett*, *arf5/mp*, and *arf7/nph4* mutants (Okushima *et al.*, 2005). *arf* mutations can affect gynoecium patterning (ARF3); impaired hypocotyl response to blue light, growth and auxin sensitivity (ARF7); the formation of vascular strands and embryo axis formation (ARF5); suppression of the hookless phenotype and hypocotyl bending (ARF2); hypocotyl elongation, and auxin homeostasis (ARF8). Some double mutants with a phenotypic single gene knock out family members may show new phenotypes. For example, the *arf7 arf19* double mutant was severely impaired in lateral root formation and showed abnormal gravitropism in both hypocotyls and roots.

To test the functional analysis of a previously cloned gene *OsARF1* we generated transgenic plants expressing an antisense (AS) copy of *OsARF1* cDNA in rice. Using a combination of morphological and molecular analysis, we show that AS-copy of *OsARF1* plants exhibits low growth and sterility.

Materials and methods

Germination assays and tissue culture

Rice callus was induced from the seeds of Zhonghua11 (*Oryza sativa L. japonica*) on MS medium that contained MS salts (Sigma, St Louis, MO, USA), 2 mg/l 2,4-D, B5 vitamins and 3% (w/v) sucrose. The primary callus was dissected from the seeds and placed on the same MS medium for 2 weeks. The subcultured callus was separated from the primary callus. After two weeks, one part of subcultured callus was placed on the same MS medium to regenerate the tertiary callus (designated as embryogenic callus) and another part was placed on the MSD medium (containing MS salt, 2 mg/l 6-BA, 0.5 mg/l kinetin, B5 vitamins and 3% sucrose), on which the callus was cultivated until a large number of green dots were growing (designated as differentiating callus). The embryogenic callus and the differentiating callus were collected at the same time for RNA extraction. The sterilized rice seeds germinated and grew to the seedling stage at 25°C under a 16-h light/8-h dark photoperiod. After the 2-leaf stage, the seedlings were transplanted into the field. The leaves and roots at the 4-leaf stage were collected. Additionally, at the reproductive stage, the young panicles that were shorter than 5 mm in length were collected and immediately frozen in liquid nitrogen and stored at −80°C.

RNA extraction and reverse transcription

Total RNA was extracted from different rice tissues by using Trizol reagent (Invitrogen, Karlsruhe, Germany) according to the manufacturer's protocol. Contaminating genomic DNA was removed by treatment with RNase-free DNase I (Sigma) at 37°C for 1 h. Total RNA amount was quantified by UV spectrophotometry at 260 nm. All the RNA samples were adjusted to the same concentration for further use. The cDNA strands were generated by reverse transcribing 5 µg of total RNA (50 µl reaction volume) using AMV reverse transcriptase (Takara Biotechnology, Shiga, Japan) at 42°C for 1 h.

Semi-quantitative RT-PCR

The cDNA samples were diluted to 1/5, 1/20 and 1/100. 1 µl of each dilution and the undiluted sample was used as template for PCR quantification. The actin specific forward primer: 5′-CCATTGGTGCTGAGCGTT-3′ and reverse primer: 5′-TAGGAATGGAAGCTGCGGG-3′ were used to quantify the amount of amplico effective cDNA template. PCR was carried out as follows; 95°C for 3 min; followed by 30 cycles of 94°C for 30 sec; 56°C for 30 sec and 72°C for 30 sec; finally a extension of 10 min. Precisely 5 µl of each amplification was loaded on 1.5% (w/v) agarose gel and the gel was stained by a 10 µg/ml ethidium bromide solution to visualize the bands. The gel was photographed by UV analysis system FR200A (Mikuni, Tokyo, Japan). The relative density of each band was measured by software Smart view™. The cDNA concentrations were normalized based on the concentration of actin. A pair of primers P-5: 5′-CTGCGTGTTGGAGTCAGGCG and P-2: 5′-CATTTGGTTCAGACTCGACCACAAC that were specific to the DNA binding domain of *OsARF1* was applied for semi-quantitative RT-PCR. Each PCR reaction was performed in a 25 µl aliquot as follows; 95°C for 3 min; followed by 30 cycles at 94°C for 30 sec; 60°C for 30 sec and 72°C for 30 sec; finally with an extension of 10 min. The following steps were the same as for actin quantification.

Construction of antisense OsARF1 transgenic rice

The original partial cDNA, corresponding to the full length coding sequence from 514 bp to 1190 bp, was inversely inserted between *Bam*HI and *Sal*I site of the binary vector pBin438. Expression was driven by Cauliflower mosaic virus (CaMV) 35S promoter and terminated by the *nos* terminator (Fig. 1A). The recombinant plasmid was transformed into rice embryogenic calli through *Agrobacterium*-mediated transformation as previously described (Sunilkumar *et al.*, 1999). Transformed calli were inculcated into selection medium supplemented by kanamycin (100 mg/l) and incubated in the dark at 28 °C for 2 weeks. Resistant-calli were sub-cultured in differentiation medium to regenerate plantlets. Primary transformants were grown in the greenhouse at 30±2 °C during the day (14 h light) and 22±2°C during the night (10 h dark). When the 4th leaf grew, the transformed plantlets were transplanted into the field for phenotypic analysis. The controls were treated under the same conditions.

Northern hybridization analysis

The whole plants at later vegetative stage were collected for RNA extractions. Total RNA (30 µg) of each sample of the transgenic plants and controls were separated by electrophoresis in agarose-formaldehyde gels and transferred to positively charged nylon membranes (Boehringer, Mannheim, Germany). Hybridization was performed with Rapidhyb™ hybridization buffer (Amersham Piscataway, NJ, USA) at 65 °C with a γ – [32P]-dCTP random-labeled probe corresponding to *OsARF1* cDNA. After washing with 2× SSC, 1× SSC and 0.5× SSC washing solutions (1× SSC is 0.15 M NaCl, 0.015 M sodium citrate) containing 0.2% (w/v) SDS at 63 °C, membranes were autoradiographed for 16–96 h using an intensifying screen (Biomax, Kodak,Rochester, NY).

Microarray analysis

Microarray analysis was carried out by Capital Biochip Corporation with hybridization tuned for gene expression variation (Lan *et al.*, 2004). The RNA samples from 11 independent transgenic plants were quantified and evenly contributed to the RNA population for microarray

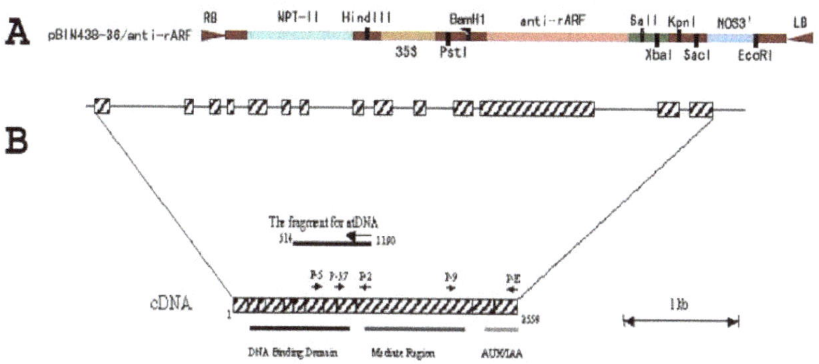

Fig. 1. **A** Construction of binary vector pBin438 harboring the AS-cDNA of *OsARF1* in the antisense orientation driven by *35S* promoter and *Nos* terminator. *Kpn*1 was within the selectable marker gene. **B** Structural features of *OsARF1* gene. The exons and introns are represented by boxes and lines, respectively.

analysis. The RNA of untransformed plants was used as the control. Probe hybridization and scanning of the hybridized microarray slide were done according to the method of Lan, (2004). Hybridization was carried out twice and data were analyzed with ArrayGuage™ version 1.21 (Fujifilm, Tokyo, Japan).

Results and discussion

Structural features of OsARF1 *gene*

A cDNA of a rice *ARF* orthologue was isolated by subtractive suppressed hybridization (SSH), using a cDNA library of young panicles as tester and that of meritstems as driver (data not shown). The 5′-end and 3′- ends were cloned using RACE method (Subudhi and Huang, 1999). By screening a bacterial artificial chromosome (BAC) library of 'IR64 *Indica*' rice variety, a BAC clone containing the genomic DNA region corresponding to the cDNA was identified. The region of approximately 8 kb that encompassed the coding region was sequenced by primer walk sequencing. A rice marker RZ797, which mapped on chromosome 11, was within this region locating *OsARF1*. The full-length cDNA of the rice *ARF1* gene was 2,559 bp; consisted of 14 exons; and encoded 852 amino acid polypeptides (Fig. 1B). Similarity searches indicated that the *OsAFR1* gene was most closely orthologous to *Arabidopsis ARF1*. Therefore, '*Oryza sativa* auxin response factor 1' (*OsARF1*; GenBank accession number AJ306306) was named according to the standard nomenclature. The deduced amino acid sequence of *OsARF1* had the typical features of transcriptional factor, with a DNA binding domain (AuxRe) close to its N- terminal, and a protein interaction domain *(*Aux/IAA*)* at its C- terminal. However, the first 50 amino acid at N-terminus and the mediate region between AuxRe and Aux/IAA showed low identities to any known ARF family gene (Fig. 1B).

Expression pattern of OsARF1 gene in specific rice tissues

The expression level of *OsARF1* was extremely low so that it was not detected by Northern hybridization; even in young panicle where the gene was originally isolated. Consequently the *OsARF1* transcript abundance patterns were analyzed using semi-quantitative RT-PCR. The 5′ primer was designed to a region of *OsARF1* sequence that was not conserved in other ARF or *Aux/IAA* genes. *OsARF1* was transcript accumulated in callus and young panicle at much higher amount than in leaf and root. Compared to that in non-differentiated callus, the transcript abundance in differentiating callus dropped significantly (Fig. 2). The results suggested that *OsARF1* could be associated with embryogenesis, because of the higher transcript abundance in young panicles and calli, which both are embryonic tissues.

AS-OsARF1 transgenic rice showed inhibited growth

To investigate the effects of the loss of function of the *OsARF1* gene on rice phenotypic characters, the *AS-OsARF1* vector was constructed and transferred to the rice cells using *Agrobacterium*-mediated transformation technique. The transgeneic plants were detected by PCR using non-intron sequence specific forward primer P-5.7: 5′-AAATGGCGCTCCCTTAAGGTGAG and the reverse primer P-2: 5′-CATTTGGTTC GACTCGACCACAAC. Eleven of 16 regenerated rice plants showed the

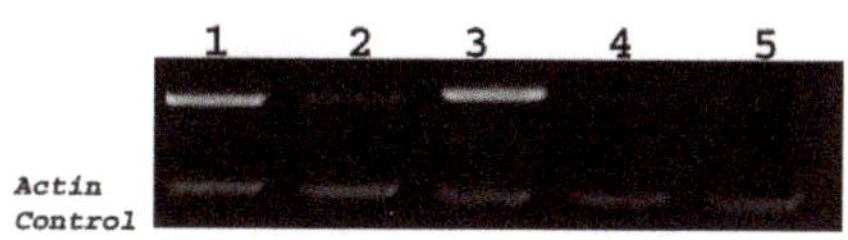

Fig. 2. Transcript abundance of *OsARF1* in wild type rice. Lanes were; 1. embryogenic callus; 2. differentiating callus; 3. young panicles; 4. leaves; 5. roots. Lower lanes: *Actin* mRNAas a control.

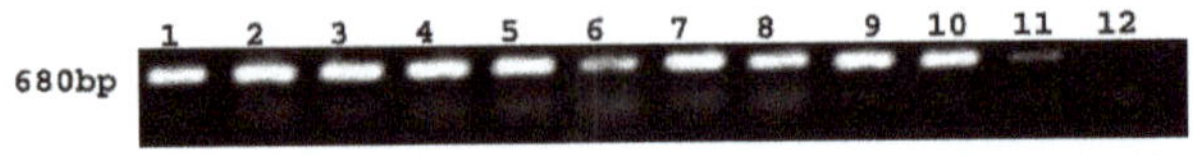

Fig. 3. Detection of *AS-OsARF1* integration by non-intron sequence specific primers. Lane 1–11: *AS-OsARF1* transformed regenerated plants; Lane 12: control plant.

expected amplicons (680 bp) while there was no signal in the control, indicating that the *AS-OsARF1* cDNA was successfully integrated into the genome of transgenic plants (Fig. 3). The *AS-OsARF1* transformed plants showed significantly lower growth and vigor that included smaller leaves and shorter heights, compared to non-transformed plants. Interestingly, the length of the tillering node didn't change significantly in transgenic plants (Fig. 4). Also, the leaves curled across the width. Eight of 11 transgenic plants failed to head and the remaining three were sterile, although they were able to develop to the heading stage 13–15 days later than the non-transformed plants. Hence, no F_2 progeny was obtained. In order to investigate the transcript abundance of *AS-OsARF1*, total RNA was isolated from transgenic and control plants and subjected to RNA gel blot analysis. The result showed that there was extremely low level of mRNA in control and transgenic plants (Fig. 5).

Microarray analysis of regulated genes expression in transgenic plants

To test the changes in gene expression caused by *OsARF1* in transgenic plants, microarray analysis was conducted. Genomic transcript abundances pattern in the transgenic plants was checked with a cDNA microarray containing 14,688 genes arrayed in duplicates that were randomly selected from rice ESTs library (http://microarray.rice.dna.affrc.go.jp). Microarray analysis of the transgenic plants in later vegetative stages showed that 10,325 genes of the total of 14,688 genes were expressed at this stage in transgenic plants (Fig. 6A). The genes whose florescent ratio was <0.5 or >2.0 (transgenic/control) were defined as differently expressed genes. By this criterion, 435 genes were differently expressed, 255 of them were down regulated (Fig. 6B) and 180 were up regulated (Fig. 6C). The annotated genes were located in 9 sub-categories of molecular functions in gene ontology (Table 1). Most of the differently expressed genes were categorized among those encoding proteins with catalytic activity. However, the inference of gene function based on homology and annotation may not reflect the real function, thus the classification of these genes was only a reference.

Fig. 4. Comparison of *AS-OsARF1* transgenic rice and a non transgeneic control. Left; *AS-OsARF1* transgenic rice. Right; untransformed control.

In this study, "knock down" phenotypic analysis was used. Phenotypes were clear and consistent but unlike those seen from ortholgues in *A. thaliana*. Semi-quantitative RT-PCR showed that no transcript of the original *OsARF1* was detected in *AS-OsARF1* transgenic plants, suggesting that the AS-RNA not only inhibited the translation of its target mRNA, but also caused degradation of target mRNA. So, AS-RNA might work in a similar way as RNAi (RNA interference). Recently, it was reported that the expression of *OsARF1* was positively regulated by auxin concentration (Waller *et al.*, 2002). However, the expression level of *ARF1* in *Arabidopsis* was relatively consistent regardless the concentration of auxin (Ulmasov *et al.*, 1997a). That suggested that the *OsARF1* might not be the exact ortholog of *AtARF1* even though these two genes shared the highest degree of similarity between rice and *Arabidopsis* (Frank *et al.*, 2002). Here, *OsARF1* was expressed significantly higher in embryonic tissues (callus and young panicle) than that in non-embryonic tissues (root, leaf and differentiating callus). Therefore the endogenous auxin to cytokinin ratio might be higher in embryonic tissues. In addition, the increasing of the expression of *OsARF1* could be essential to embryogenesis since all the *AS-OsARF1*transgenic plants were sterile. *OsARF1*, as an early response gene in auxin signaling (Waller *et al.*, 2002) might play a key role in fertility and regulate quite a few downstream genes associated with growth and reproduction.

The cDNA microarray provided an efficient high throughput approach to investigate the function of a large numbers of genes and their expression in response to environmental effects (Kawasaki *et al.*, 2001). The results

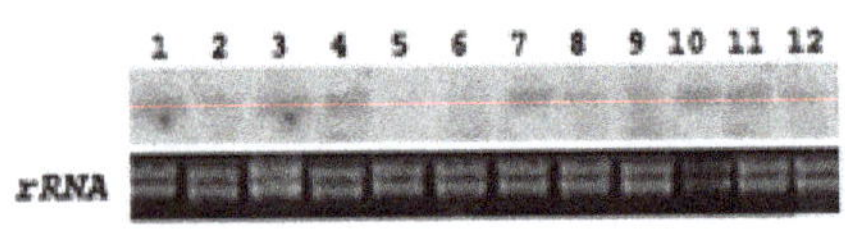

Fig. 5. Abundances of the *AS-OsARF1* mRNA in transgenic rice plants. Upper panel: Lane 1: control; lanes 2–12: 11 transgenic plants. Lower panel: rRNA control.

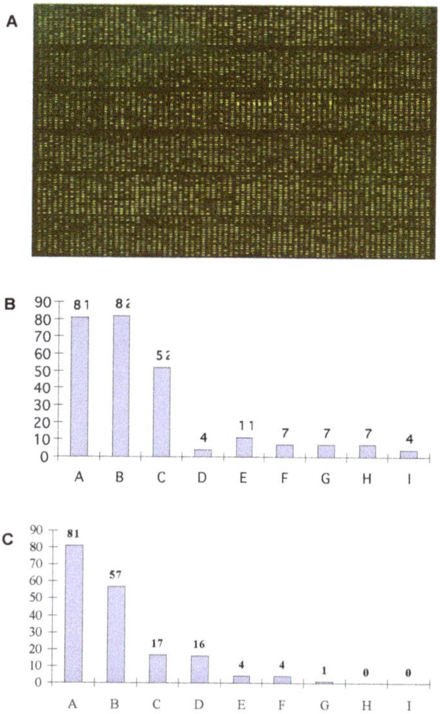

Fig. 6A. Microarray analysis of regulated transcript abundance among transgenic plants. Three-color image of a transgenic plant cDNA array. After scanning for each fluorescent dye separately false colour images were superimposed. Red dots correspond to genes higher expressed down-regulation, green dots to genes that are higher expressed up-regulation, and yellow dots represent genes that show no differential expression. **B** Genes decreased in TA in transgenic rice. Alphabetic letters represent indicators in Table 1. **C** Genes increased in TA in transgenic rice. Alphabetic letters represent indicators in Table 1.

here indicated that approximately 3% of the total set of predicted genes and approximately 4% of known to be expressed genes showed significant differences among their transcriptional abundance between the transgenic plants and the controls. The ratio of differently expressed genes was similar among different tissues. Reproductive stage analyses were not possible because all AS plants failed to reach to productive growth stage. It seems likely the knock down of *OsARF1* turned off some key switches for transformation from vegetative stage to the reproductive stage. Perhaps *OsARF1* itself was the switch. Suppression of *OsARF1* expression also caused dramatically growth at vegetative stages, even though its expression was extremely low at this stage in wild type rice. This demonstrates that, perhaps *OsARF1* is crucial to all growth. In comparison the phenotypic effect of *ARF1* in *Arabidopsis* is unknown (Wang *et al.*, 2007). In contrast, *OsARF1* was shown to be essential for the developmental growth stages of the life cycle of the rice plant. The pathways and functional mechanisms will be studied further.

References

Akila, O., Danielle, C., Alain, D.M., Yves, D., Jean, B., and Mercè, B. (2004) The T-DNA ORF8 of the cucumopine-type *Agrobacterium rhizogenes* Ri plasmid is involved in auxin response in transgenic tobacco. Plant Sci. *166*, 557–567.

Wang, D., Kemei, P., Yaping, F., and Zongxiu, S. (2007) Genome-wide analysis of the auxin response factors (ARF) gene family in rice (*Oryza sativa*) Gene *394*, 13–24.

Frank, W., Masaki, F., and Peter. N. (2002) *OsARF1*, an auxin response factor from rice, is auxin-regulated and classifies as a primary auxin responsive gene. Plant Mol. Biol. *50*, 415–425.

Kawasaki, S., Borchert, C., Deyholos, M., Wang, H., Brazilla, S., Kawai, K., Galbraith, D., and Bohnert, H. (2001) Gene expression profile during the initial phase of salt stress in rice. Plant Cell *13*, 889–905.

Lan, L., Chen, W., Lai, Y., Suo, J., Kong, Z., Li, C., Lu, Y., and Zhang, Y. (2004) Monitoring of gene expression profiles and isolation of candidate genes involved in

Table 1. Genes with different transcript abundances detected by cDNA microarray.

	Differently expressed genes			
No.	Categorization of expressed genes	Down-regulated genes	Up-regulated genes	Total numbers of expressed genes
A	No significant homology in GenBank	81	81	162
B	Homology to function unknown genes	82	57	139
C	Catalytic activity	52	17	69
D	Structural molecule activity	4	16	20
E	Transporter activity	11	4	15
F	Binding	7	4	11
G	Antioxidant activity	7	1	8
H	Enzyme regulator activity	7	0	7
I	Signal transducer activity	4	0	4
Total		**255**	**180**	**435**

pollination and fertilization in rice (*Oryza sativa L.*) with a 10K cDNA microarray. Plant Mol. Biol. *54*, 471–87.

Okushima, Y., Overvoorde, P.J., Arima, K., Alonso, J.M., and Chan, A. (2005) Functional Genomic Analysis of the AUXIN RESPONSE FACTOR Gene Family Members in *Arabidopsis thaliana*: Unique and Overlapping Functions of ARF7 and ARF19. Plant Cell *17*, 444–63.

Ozga, J.A., Huizen, R., and Reinecke, D.M. (2002) Hormone and seed-specific regulation of pea fruit growth. Plant Physiol. *128*, 1379–89.

Remington, D.L., Visionm, T.J., Guilfoylem, T.J., and Reed, J.W. (2004) Contrasting models of diversification in the *Aux/IAA* and *ARF* gene families. Plant Physiol. *135*, 1738–1752.

Subudhi, P.K., and Huang, N. (1999) RAPD mapping in a doubled haploid population of rice (*Oryza sativa L.*) Hereditas *130*, 41–9.

Sunilkumar, G., Vijayachandra, K., and Veluthambi, K. (1999) Pre-incubation of cut tobacco leaf explants promotes *Agrobacterium*-mediated transformation by increasing *vir* gene induction. Plant Sci. *141*, 51–58.

Ulmasov, T., Liu, Z.B., Hagen, G., and Guilfoyle, T.J. (1995) Composite structure of auxin response elements. Plant Cell *7*, 1611–1623.

Ulmasov, T., Hagen, G., and Guilfoyle, T.J. (1997a) ARF1, a transcription factor that binds to auxin response elements. Sci. *276*, 1865–1868.

Ulmasov, T., Hagen, G., and Guilfoyle, T.J. (1999) Activation and repression of transcription by auxin-response factors. PNAS *96*, 5844 – 5849.

Ulmasov, T., Hagen, G., and Guilfoyle, T.J. (1999a) Dimerization and DNA binding of auxin response factors. Plant J. *19*, 309–319.

Waller, F., Furuya, M., and Nick, P. (2002) OsARF1, an auxin response factor from rice, is auxin-regulated and classifies as a primary auxin responsive gene. Plant Mol. Biol. *50*, 415–25.

Chapter 5

An Account of Cloned Genes of Methyl-erythritol-4-phosphate Pathway of Isoprenoid Biosynthesis in Plants

Deepak Ganjewala, Shiv Kumar, and Rajesh Luthra

Abstract

Isoprenoids, also known as terpenoids, are biosynthesized by the condensation of the two C_5 unit isopentenyl diphosphate (IPP) and isomer dimethylallyl diphosphate (DMAPP). Generally, plants use two separate pathways plastidial Methyl-erythritol-4-phosphate (MEP) and cytosolic acetate-mevalonate (MVA) pathways for formation of IPP. The genes, enzymes and intermediates of the MEP pathway have been unravelled in plants over the past few years. Interestingly, MEP pathway enzymes are encoded by nuclear genes but they function in plastids to produce precursors for isoprenes, monoterpenes, carotenoids, abscisic acid, gibberellins, and the side chain of chlorophylls, tocopherols, phylloquinones, and plastoquinone. In *Arabidopsis thaliana*, a complete set of genes of MEP pathway homologous to the *E. coli* MEP pathway genes have been identified. Although, these genes have been cloned and characterized from several other plants but overall information about them at one place is not available so far. Though, a range of reviews are available about their roles in isoprenoid biosynthesis and regulation. Therefore, we decided to compile the data on cloned and characterized genes of MEP pathway in plants. Also, we summarize the results of the previously published reports, particularly those which were based on incorporation of ^{13}C-glucose or by application of specific inhibitors such as mevinolin and fosmidomycin to look into the MEP pathway in plants. In addition, we searched for the two key enzymes DXS and HMGR that could be assigned for the acetate-MVA and MEP pathway with the help of bioinformatics tools. Presence or absence of these enzymes can be correlated with respective isoprenoid biosynthetic pathways in plants.

Introduction

Plant secondary metabolites *viz.*, isoprenoids have always been fascinating for the researchers. Obviously, they are the largest and most diverse class of plant secondary metabolites (Dubey *et al.*, 2003). Over the years isoprenoids have been exhaustively investigated. There are many research papers, reviews, commentaries and books are available covering various aspects of isoprenoids *viz.*, biosynthesis and regulation and functions in plants and microorganisms (Croteau, 1987; Rohmer, 1999; 2003; Lichtenthaler, 1999; Lichtenthaler, 2001; Eisenreich *et al.*, 1997; 2001; 2004; Rodrıguez-Concepcion and Boronat, 2002; Chemler *et al.*, 2006; Cheng *et al.*, 2007). Isoprenoids are multifunctional; they play very important roles in membrane structure, redox reactions, light harvesting and photo-protection, and regulations of growth and development. They not only perform various roles in the plant's life such as, plant-environment, plant–insect, plant–microorganism and plant–plant interactions but they became an essential part of our life as medicines, flavours, fragrances, cosmetics, dyes, insecticides and more (Verpoorte *et al.*, 2002; Harborne, 2001; Dixon, 2001). Currently, isoprenoids are being used as anti-cancer and antimicrobial drugs for example artimicinin as a powerful antimalarial (Dhingra *et al.*, 2000) and taxol as anti-cancer (Cragg *et al.*, 1997) agents.

Despite the great deal of structural and functional diversity all isoprenoids are synthesized by consecutive condensation of common C_5 isoprene precursor; isopentenyl diphosphate (IPP) and its isomer dimethylallyl diphosphate (DMAPP). IPP therefore is regarded as the universal precursor of all the isoprenoids. IPP, in turn is biosynthesized in two different sites *via* two separate and independent biochemical pathways: 1. Cytosolic acetate-MVA pathway and 2. Plastidial Methylerythritol 4- phosphate (MEP) also called as 1-deoxy-D-xylulose 5-phosphate (DOXP) or glyceraldehyde-3-phosphate-pyruvate (GAP-Pyruvate) pathway (Fig. 1). In the present article, the pathway is described as MEP pathway. In general, acetate-MVA pathway produces sesquiterpenes and triterpenes while MEP pathway produces monoterpenes, sesquiterpenes, diterpenes, tetraterpenes, plastoquinone and prenyl side chains of chlorophyll.

In brief, MEP pathway (Fig. 1) begins with the formation of 1-deoxy-D-xylulose 5-phosphate (DOXP/DXP) by the condensation of pyruvate and glyceraldehyde 3-phosphate catalysed by DOXP synthase (DXS,

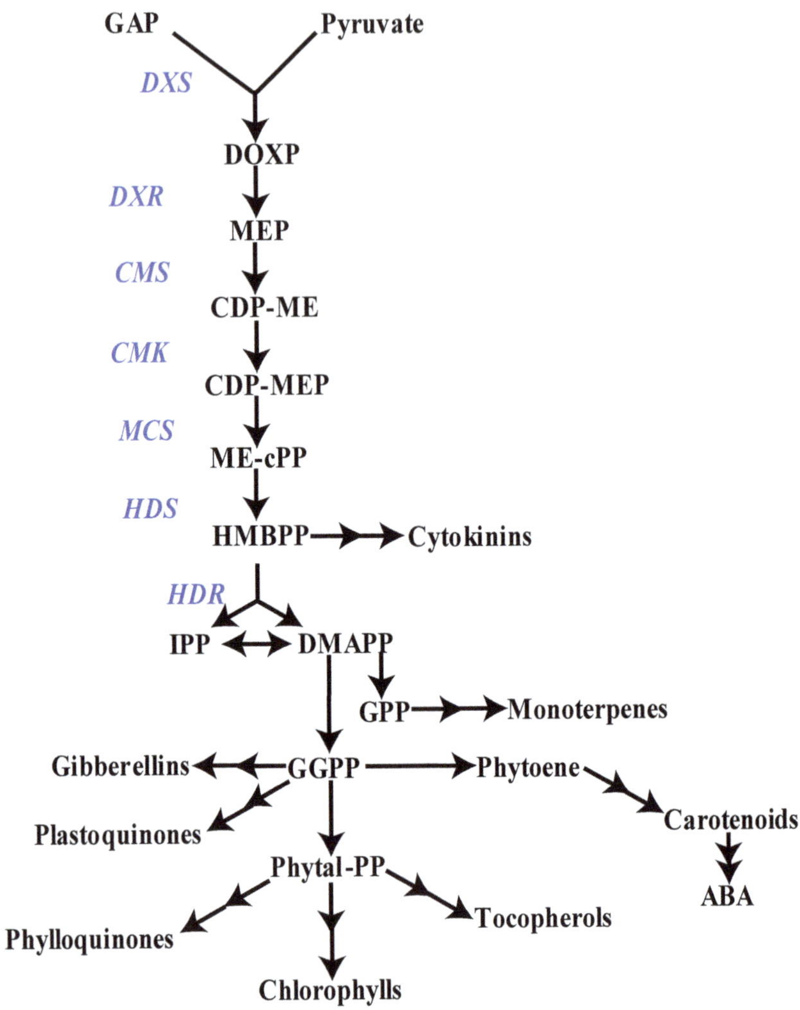

Fig. 1. MEP pathway for the biosynthesis of isoprenoids in plants. ABA, abscisic acid; **CDP-ME, 4-(cytidine-5'-diphospho)-2-C-methyl-D-Erythritol**; **CDP-MEP, 4-Diphosphocytidyl-2C-methyl-D-erythritol 4-phosphate**; DMAPP, dimethylallyl diphosphate; **DXP, 1-deoxy-D-xylulose 5-phosphate**; GAP, glyceraldehyde 3-phosphate; GGPP, geranylgeranyl diphosphate; GPP, geranyl diphosphate; **HBMPP, 4-hydroxy-3-methylbut-2-enyl diphosphate**; IPP, isopentenyl diphosphate; **ME-cPP, 2C-methyl-D-erythritol 2,4-cyclodiphosphate**; **MEP, 2C-methylerythritol 4-phosphate**. Enzymes are indicated in bold italic; 4-(cytidine-5'-diphospho)-2-C-methyl-D-Erythritol kinase **(CMK, EC 2.7.1.148)**; 4-Diphosphocytidyl-2C-methyl-D-erythritol 4-phosphate synthase **(CMS, EC 2.7.7.60)**; 1-deoxy-D-xylulose 5-phosphate reductoisomerase **(EC 1.1.1.267)**; 1-deoxy-D-xylulose 5-phosphate synthase **(DXS, EC 4.1.3.37)**; 4-hydroxy-3-methylbut-2-enyl diphosphate reductase **(HDR, EC 1.17.1.2)** ; 4-Hydroxy-3-methylbut2-en-yl-diphosphate synthase **(HDS, EC, 1.17.4.3)**; 2C-methyl-D-erythritol 2,4-cyclodiphosphate synthase **(MCS, EC 4.6.1.12)**.

(EC 4.1.3.37). DOXP then undergo intra-molecular rearrangement and reduction catalysed by 1-deoxy-D-xylulose 5-phosphate reductoisomerase (DXR, **EC 1.1.1.267)** to yield methyl erythritol-4-phosphate (MEP) which is regarded as an immediate precursor of plastidic isoprenoids. MEP is consecutively converted into 4-diphosphocytidyl-methylerythritol (CDP-ME), 4-diphosphocytidyl-methylerythritol (CDP-MEP) and methyl-erythritol 2,4-cyclodiphosphate (ME-cPP). These reactions are carried out by CDP-ME synthase (CMS, **EC 2.7.7.60**), CDP-ME kinase (CMK, **EC 2.7.1.148**) and ME-cPP synthase (MCS, **EC 4.6.1.12**). Methyl-erythritol 2,4-cyclodiphosphate (ME-cPP) then converted to hydroxymethylbutenyl 4-diphosphate (HMBPP) by an enzyme hydroxymethylbutenyl 4-diphosphate synthase (HDS, **EC 1.17.4.3**). HMBPP is finally converted into a mixture of IPP and DMAPP by the enzyme HMBPP reductase (HDR, **1.17.1.2**) (Eisenreich *et al.*, 2001; 2004).

At present, the genes encoding enzymes of MEP pathway with homology to the *E. coli* MEP pathway enzymes haven been identified from variety of plants including *Arabdopsis thaliana* (http://www.Arabidopsis.org). The present review is aimed to provide the information on these cloned and characterized genes of MEP pathway in plants. Basically, the idea to compile the information on MEP pathway genes came from a very likely article on cloned genes of acetate-MVA pathway in plants by Scolink and Bartley (1996). The information about cloned genes of MEP pathway in one piece would be helpful to home in to isoprenoid biosynthesis. The results presented here were derived through online search and analysis (http://www.ncbi.nlm.nih.gov/; http://www.tigr.org/). In addition, published literatures were also taken in account such as those describe application of ^{13}C-glucose-Nuclear Magnetic Resonance (NMR) spectroscopy for elucidation of MEP pathway in plants. In addition, we took the help of bioinformatics tools to search DXS and HMGR, two key enzymes of their respective pathways to investigate these pathways in plants.

^{13}C-glucose-NMR spectroscopy provides clues for the MEP pathway

The potential of using ^{13}C-glucose-NMR spectroscopy to elucidate metabolic pathways in plants has long been recognized. Early efforts relied on NMR spectra of metabolites which were related to the underlying pathways used to create them (Jeffrey *et al.*, 1991). NMR spectra have also been used to elucidate the flux through metabolic pathways (Bacher *et al.*, 1998; Kelleher, 2001). The use of *in vivo* ^{13}C-NMR spectroscopy to study the biosynthesis of secondary metabolites in plants has been well documented previously (see Table 2). Useful clues to the origin of the carbon atoms of isoprenoids in bacteria and for the elucidation of the MVA independent route were obtained from labeling experiments using ^{13}C-glucose with *Zymomonas mobilis*, a facultative anaerobic and fermentative bacterium (Sprenger, 1996). The use of NMR spectroscopy in plant secondary metabolism has been hampered mainly because of very low concentration of the secondary metabolites, and the pathway leading to isoprenoid formation are often branched hence the ^{13}C-label of early precursors is diluted into several metabolites at the end. These difficulties in resolving the origin of Isoprenoid units could be overcome by NMR analysis of extracts or isolated compounds. To investigate the biosynthetic origin of isoprenoid building blocks of secondary metabolites, the pathway-independent precursor ^{13}C-glucose, which produces distinctly different labelling patterns of the individual isoprene units for the MEP and MVA pathways is generally employed (Rohmer, 1999). Given that glucose is a general intermediary metabolite, the isotope from the proffered carbohydrate can be diverted to virtually all metabolic compartments and intermediates in plant cells (Eisenreich *et al.*, 2004).

The biosynthetic origin of a considerable number of primary and secondary plant terpenoids has been and currently being reinvestigated using ^{13}C-glucose-NMR spectroscopic technique in higher and lower plants (liverworts). Table 1 provides the information on application of ^{13}C-NMR spectroscopy to elucidate the MEP pathway in plants. The data show that a wide variety of monoterpenes, diterpenes and sesquiterpenes (germacrene) are biosynthesized predominantly *via* the MEP pathway. Beside the analysis of these published reports on MEP pathways, our online database search for DXS and HMGR the key regulatory enzyme respectively of MEP and acetate-MVA pathways further provided the clue for the operation of either of these pathways in plants. Table 2 provides information on distribution of DXS and HMGR enzymes in plants. This information was collected online from http://www.ncbi.nlm.nih.gov/.

Cloned genes of MEP pathway

Our knowledge and understanding about the biosynthesis and regulation of isoprenoids in plants has been tremendously increased during the past two decades. As a result, genes encoding enzymes of the MEP pathway have been cloned and characterized from a several plants in the recent time. Though, several genes of the MEP pathway downstream from *ispC* were discovered by a strategy combining biochemical evidence with comparative genomic analysis. Please see a review by Eisenreich *et al.* (2004) for detailed description about the mechanism of action of enzymes of MEP pathway. Here in the Table 3 we provide information exclusive on cloned and characterized genes of the MEP pathway in plants. The information mainly comprises of GeneBank accession number, size and protein or gene name. Table 3 shows that DXS and DXR have been cloned and characterized from a variety of plants while other enzymes of the MEP pathway could only be characterized only from a few plants. So far, HDS is known from only two plants *Nicotiana benthamiana* and *Oryza sativa*. Similarly HDR is also known from only two plants, *Arabidopsis thaliana* and *N. benthamiana*. An overall distribution of MEP genes in plants is presented in Table 4. From Table 4, it is clear that *A. thaliana* and *O. sativa* (Japonica var.) genome has complete set of genes encoding enzymes of MEP pathway along with HMGR of acetate-MVA pathway. *Stevia rebaudiana* genome has shown at least six of the seven genes of MEP pathway, but lacks the HMGR of acetate-MVA pathway. Further analysis of the data has revealed that 19 out of 39 plants searched online have

Table 1. MEP pathway in plants verified on the basis of ^{13}C-glucose-NMR spectroscopy.

Plants	Isoprenoids	References
Catharanthus roseus	Terpenoids Iridoid glucoside secologanin	Arigoni *et al.*, 1997; 1999; Contin *et al.*, 1998
Chelidonium majus *Populus nigra* *Salix viminalis*	Isoprene	Zeidler *et al.*, 1997, 1998
Conocephalum conicum	Isoprenoids	Thiel *et al.*, 2002
Dacus carota *Hordeum vulgare* *Lemna gibba*	L-carotene, lutein, prenyl chains of chlorophylls and plastoquinone-9	Lichtenthaler *et al.*, 1997
Eucalyptus globules	Cineol	Rieder *et al.*, 2000
Fossombronia alaskana	Hopane triterpene and three diterpenes	Hertewich *et al.*, 2001
Hordeum vulgare	sesquiterpenoid cyclohexane derivatives	Maier *et al.*, 1998
Liriodendron tulipifera	Terpenes	Sagner *et al.*, 1998
Marrubium vulgare	Labdane diterpenoid marrubin	Knoss *et al.*, 1997
Matricaria recutita	Isoprene units of chamomile sesquiterpenes	Adam and Zapp, 1998
Mentha citrate	Linalyl Acetate monoterpene	Fowler *et al.*, 1999
Mentha pulegium	Monoterpenes	Eisenreich *et al.*, 1997
Narcissus pseudonarcissus	*b*-carotene	Fellermeier *et al.*, 1999
Pelargonium graveolens *Thymus vulgaris*	Monoterpenes	Eisenreich *et al.*, 1997
Persea Americana	Abscisic acid; Carotenoids and abscisic	Hirai *et al.*, 2000; Milborrow *et al.*, 1998
Rauwolfia serpentine	Monoterpene loganin	Eichinger *et al.*, 1996
Taxus chinensis	Taxol (diterpene)	Eisenreich *et al.*, 1996
Trichcolea tometella	Trichcolein and deoxytometellin (Hemi- and mono- terpene moieties and diterpene phytol)	Barlowa *et al.*, 2003
Vitis Vinifera	Linalool and geraniol	Klink *et al.*, 2005
Lepidolaena hodgsoniae	Sesquiterpene hodgsonox	Luan *et al.*, 2002
Anisotome layallii	Anisotomenes (bicyclic irregular diterpenes)	Barlowa *et al.*, 2003
Piper aduncum	Isoprene Units in Chromenes	Leite *et al.*, 2007
Solidago Canadensis	Germacrene D (sesquiterpenes)	Steliopoulos *et al.*, 2002

genes of both MEP and acetate-MVA pathway in their genome, while the others 20 plants had exclusively genes of MEP pathway.

Conclusion

Many pathogenic microorganisms including *Mycobacterium tuberculosis* and *Plasmodium falciparum* also operate MEP pathway for the biosynthesis of isoprenoids. In fact, isoprenoids plays crucial role in the survival of *P. falciparum* in host cells. Knowledge of the MEP pathway in such pathogenic microorganism is currently being exploited for the development of structure-based anti-microbial drugs by targeting the enzymes of MEP pathway. Therefore, details concerning the genes, enzymes and intermediates of the MEP pathway have become essential in achieving these goals. Currently, fosmidomycin an inhibitor of DOXP reductoisomerase (DXR) of MEP pathway has been successfully tested to hang-up isoprenoid biosynthesis in *P. falciparum*. Similar strategies could be employed for the development of novel herbicides (Lichtenthaler *et al.*, 2000). This aspect of the isoprenoids researches have a direct impact on human health, hence created much interest and awareness among the researchers in the recent years to look for new structure based drugs against more pathogenic microorganisms and weeds relying on MEP pathway. Our, knowledge and understanding about the plant secondary metabolite biosynthesis and regulation has greatly accelerated these efforts. Most certainly, comparative genomics and in combination of bioinformatics has been an aid.

Acknowledgements

Authors are grateful to the Chancellor, VIT University for providing the necessary support and Professor Lazar Mathew and Professor Braj Nandan Prasad, School of Biotechnology, Chemical and Biomedical Engineering, VIT University for their constant encouragement and valuable suggestions. Financial support from the Department of Sciences and Technology (DST) and Council of Scientific and Industrial Research (CSIR), New Delhi, Government of India is duly acknowledged.

Table 2. Results of online search for DXS and HMGR. DXS is assigned to MEP whereas HMGR to acetate-MVA pathway.

	MEP Pathway DOXP Synthase (DXS)		Acetate-Mevalonate Pathway HMG-CoA Reductase (HMGR)	
Plants	Accession No.	Size	Accession No.	Size
Arabidopsis thaliana	BAB02345	604	AAA67317	562
Antirrhinum majus	AAW28999	733		
Artemisia annua	AAD56390	713	AAD47596	567
Andrographis paniculata	AAP14353	691	AAP14352	556
Camptotheca acuminate			AAB69726	575
Capsicum annuum	CAA75778	719	AAD28179	604
Catharanthus roseus	CAA09804	716	AAT52222	601
Cistus incanus subsp. creticus			ABL10110	388
Chrysanthemum x morifolium	BAE79547	669		
Elaeis guineensis	AAS99588	707		
Ginkgo biloba	AAS89341	717	AAU89123	571
Hevea brasiliensis	AAS94123	720	CAA38467	575
Lycopersicon esculentum	AAD38941	719	AAB62581	601
Lycopersicon hirsutum	AAT97962	714		
Mentha x piperita	AAC33513	724		
Medicago truncatula	ABP03805	711	ABE88827	583
	ABP03804	710		
	ABO82094	717		
Morinda citrifolia	AAL32062	722		
Narcissus pseudonarcissus	CAC08458	709		
Nicotiana tabacum			AAB87727	604
			AAO85554	604
Oryza sativa (Japonica)	NP_001055524	720	BAD10066	561
Oryza sativa (Indiaca)	EAY98024	710	CAA92821	576
	AAB88295	594		
Pueraria montana var. lobata	AAQ84169	717		
Picrorhiza kurrooa			ABC74565	561
Salvia miltiorrhiza			ABB45812	174
			AAU87798	267
Stevia rebaudiana	CAD22155	715		
Tagetes erecta	AAG10432	725	AAC15475	574
Taxus cuspidate				
Taxus x media	AAS89342	742	AAQ82685	595
Zea mays	ABP88134	719	CAA70440	579
	ABP88135	705		
	AAX49359	481		
	AAX49358	424		
Vitis Vinifera	CAN71054	1638	CAN72217	575

DOXP: 1-deoxy-D-xylulose 5-phosphate; HMG-CoA 3-hydroxy-3-methylglutaryl-coenzyme A.

Table 3. Cloned and characterized enzymes of the MEP pathway in plants.

Enzyme	Plant	Accession numbers	Size	Protein / Gene	Reference
1-deoxy-D-xylulose 5-phosphate synthase (DXS, EC 4.1.3.37)	*Arabidopsis thaliana*	NP_001078570	565	DXPS3	Sato *et al.*, 2000
		NP_196699	700	DXPS3	
		NP_850620	629	DXPS1	
		BAB02345	604	DXS	
	Antirrhinum majus	AAW28999	733	DXPS	Dudareva *et al.*, 2005
	Capsicum annuum	CAA75778.1	719	dxs put.	Bouvier *et al.*, 1996
	Catharanthus roseus	CAA09804	716	DXS	Chahed *et al.*, 1997
	Chrysanthemum x morifolium	BAE79547	669	DXS	Kishimoto *et al.*, 2006
	Croton stellatopilosus	BAF75640	720	dxs	Wungsintaweekul *et al.*, 2008
	Elaeis guineensis	AAS99588	707	dxs	Khemvong *et al.*, 2005
	Ginkgo biloba	AAS89341	711		Gong *et al.*, 2006
	Hevea brasiliensis	AAS94123	720	DXS	Seetang-Nun *et al.*, 2008a
		ABF18929	711	DXS2	
	Lycopersicon esculentum	AAD38941	719	dxs	Lois *et al.*, 2000
	Morinda citrifolia	AAL32062	722	DXS	Han *et al.*, 2003
	Oryza sativa (Japonica)	NP_001055524	720	dxs	Ohyanagi *et al.*, 2006
	Oryza sativa (Indiaca)	AAB88295	594	CLA1	Campos *et al.*, 1997
		TC263109		DXS1	Kim *et al.*, 2005
		TC262788		DXS2	
		TC276717		DXS3	
	Picea abies	ABS50520	746	DXS2B	Phillips *et al.*, 2007
		ABS50519	740	DXS2A	
		ABS50518	717	DXS1	
	Stevia rebaudiana	CAD22155	715	dxs	Totte *et al.*, 2003
	Tagetes erecta	AAG10432	725	dxs	Moehs *et al.*, 2001
1-deoxy-D-xylulose 5-phosphate reductase (DXR, EC 1.1.1.267)	*Arabidopsis thaliana*	CAB43344	406	dxr	Schwender *et al.*, 1999
	Antirrhinum majus	AAW28998	471	dxr	Dudareva *et al.*, 2005
	Camptotheca acuminate	ABC86579	472	Dxr	Yao *et al.*, 2008
	Catharanthus roseus	AAF65154	474	dxr	Veau *et al.*, 2000
	Chrysanthemum x morifolium	BAE79548	487	DXR	Kishimoto *et al.*, 2006
	Ginkgo biloba	AAR95700	477	Dxr	Gong *et al.*, 2005
	Hevea brasiliensis	AAS94121	471	DXR	Seetang-Nun *et al.*, 2008b
	Hordeum vulgare	CAE47438	484	dxr	Hans *et al.*, 2005
	Lycopersicon esculentum	AAK96063	475	DXR	Rodriguez-Concepcion *et al.*, 2001
	Mentha x piperita	AAD24768	470	DXR	Lange and Croteau, 1999
	Oryza sativa (Japonica)	NP_001041780	473	dxr	Ohyanagi *et al.*, 2006
	Oryza sativa (Indiaca)	EAY72208	473	dxr	Yu *et al.*, 2005
	Plectranthus barbatus	AAR99081	469	dxr	Engprasert *et al.*, 2005
	Pueraria montana var. lobata	AAQ84168	465	dxr	Sharkey *et al.*, 2005
	Rauvolfia verticillata	AAY87151	474	DXR	Wu *et al.*, (in Press)
	Salvia miltiorrhiza	ABJ80680	474	DXR	Liao *et al.*, 2007
	Stevia rebaudiana	CAD22156	473	dxr	Totte *et al.*, 2003
	Taxus cuspidate	AAT47184	477	dxr	Jennewein *et al.*, 2004
	Stevia rebaudiana	CAD22156	473	dxr	Totte *et al.*, 2003
	Taxus cuspidate	AAT47184	477	dxr	Jennewein *et al.*, 2004

Table 3. Cloned

Enzyme	Plant	Accession numbers	Size	Protein / Gene	Reference
4-Diphosphocytidyl-2C-methyl-D-erythritol 4-phosphate synthase (CMS, EC 2.7.7.60)	*Arabidopsis thaliana*	NP_565286	302	ISPD	Seki *et al.*, 2002
		BAC42737	302	ispD	
	Ginkgo biloba	AAZ80386	327	MECT	Kim *et al.*, 2005
	Oryza sativa (Japonica)	BAD82130	297	ispD put.	Sasaki and Matsumoto, 2002; Yu *et al.*, 2005
	Oryza sativa (Indiaca)	EAY76759	408	ispD pat.	
4-(cytidine-5'-diphospho)-2-C-methyl-D-Erythritol kinase (CMK, EC 2.7.1.148)	*Arabidopsis thaliana*	O81014	383	ISPE	Lin and Kaul, 1999
	Lycopersicon esculentum	AAF87717	401	ispE	Lange and Croteau, 1999
	Mentha x piperita	P56848	405	ISPE	Rohdich *et al.*, 2000
	Oryza sativa (Japonica)	NP_001044544	401	ispE	Ohyanagi *et al.*, 2006
2C-methyl-D-erythritol 2,4-cyclodi-phosphate synthase (MCS, EC 4.6.1.12)	*Arabidopsis thaliana*	AAM62786	231	MECDP_S	Gao *et al.*, 2006
	Ginkgo biloba.	AAY40863	239	Mecps	Alexandrov *et al.*, 2006
	Oryza sativa (Japonica)	EAZ24186	222	MECDP_S	Yu *et al.*, 2005
	Oryza sativa (Indica)	EAY87077	222	MECDP_S	Yu *et al.*, 2005
	Taxus x media	ABB88956	247	mecs	Jin *et al.*, 2006
4-Hydroxy-3-methylbut2-en-yl-diphosphate synthase (HDS, EC, 1.17.4.3)	*Nicotiana benthamiana*	AAS75817	268	gcpE/ispG	Page *et al.*, 2003
	Oryza sativa (Japonica)	AAO72576	608	gcpE	Cooper *et al.*, 2003
1-Hydroxy-2-methyl-butenyl 4-diphosphate reductase (HDR, EC 1.17.1.2) Or, 4-hydroxy-3-methylbut-2-enyl diphosphate reductase	*Arabidopsis thaliana*	AAW82381	468	HDR/ISPH	Guevara-Garcia *et al.*, 2005
	Nicotiana benthamiana	AAS75818	166	ispH LytB	Page *et al.*, 2003

References

Adam, K.P. and Zapp, J. (1998) Biosynthesis of the isoprene units of chamomile sesquiterpenes. Phytochemistry *48*, 953–959.

Alexandrov, N.N., Troukhan. M.E., Brover, V.V., Tatarinova, T., Flavell, R.B., and Feldmann, K.A. (2006) Features of Arabidopsis genes and genome discovered using full-length cDNAs. Plant Mol. Biol. *60*, 71–87, 2006.

Arigoni, D., Eisenreich, W., Latzel, C., Sagner, S., Radykewicz, T., Zenk, M.H., and Bacher, A. (1999) Dimethylallyl pyrophosphate is not the committed precursor of isopentenyl pyrophosphate during terpenoid biosynthesis from 1-deoxyxylulose in higher plants. Proc. Natl. Acad. Sci. U.S.A. *96*, 1309–1314.

Arigoni, D., Sagner, S., Latzel, C., Eisenreich, W., Bacher, A., and Zenk, M.H. (1997) Terpenoid biosynthesis from 1-deoxy-Dxylulose in higher plants by intramolecular skeletal rearrangement. Proc. Natl. Acad. Sci. U.S.A. *94*, 10600–10605.

Bacher, A., Rieder, C., Eichinger, D., Arigoni, D., Fuchs, G., Eisenreich, W. (1998) Elucidation of novel biosynthetic pathways and metabolite flux patterns by retrobiosynthetic NMR analysis. FEMS Microbiol. Rev. *22*, 567–598.

Barlowa, A.J., Lorimer, S.D., Morgan, E.R., and Weavers, R.T. (2003) Biosynthesis of the sesquiterpene hodgsonox from the New Zealand liverwort *Lepidolaena hodgsoniae*. Phytochemistry *63*, 25–29.

Bouvier, F., d'Harlingue, A., Suire, C., Backhaus, R.A., and Camara, B. (1998) Dedicated roles of plastid transketolases during the early onset of isoprenoid biogenesis in pepper fruits. Plant Physiol. *117*, 1423–1431.

Campos, N., Lois, L.M., and Boronat, A. (1997) Nucleotide sequence of a rice cDNA encoding a transketolase-like protein homologous to the Arabidopsis CLA1 gene product. Plant Physiol. *115*, 1289–1293.

Chahed, K., Oudin, A., Guivarch, N., Hamdi, S., Chenieux, J.C., Rideau, M., and Clastre, M. (2000) 1-Deoxy-D-xylulose 5-phosphate synthase from periwinkle: cDNA identification and induced gene expression in terpenoidindole alkaloid-producing cells. Plant Physiol. Biochem. *38*, 559–566.

Chemler, J.A., Yan, Y., and Koffas, M.K.G. (2006) Biosynthesis of isoprenoids, polyunsaturated fatty acids and flavonoids in *Saccharomyces cerevisiae*. Microb. Cell Fact. *5*, 20–30.

Cheng, A.X., Lou, Y.G., Mao,Y.B., Lu, S., Wang, L.J., and Chen, X.Y. (2007) Plant Terpenoids: Biosynthesis and Ecological Functions. J. Integrative Plant Biol. *49*, 179–186.

Contin, A., Vander-Heijden, R., Lefeber, A.W.M., and Verpoorte, R. (1998) The iridoid glucoside secologanin is derived from the novel triose-phosphate/pyruvate pathway in *Catharanthus roseus* cell culture. FEBS Lett. *434*, 413–416.

Cooper, B., Clarke, J., Budworth, P., Kreps, J., Hutchison, D., Park, S., Guimil, S., Dunn, M., Luginbuhl, P., Ellero, C., Goff, S.A., and Glazebrook, J. (2003) A network of rice genes associated with stress response and seed development. Proc. Natl. Acad. Sci. U.S.A. *100*, 4945–4950.

Table 4: Overall distribution of MEP pathway genes in plants.

Plants	MEP pathway genes							
	dxs	*yaeM/dxr*	*ygbP/mect/ispD*	*ychB/cmek/ispE*	*ygbB/mecs*	*gcpE/ispG*	*hdr/ ispH*	*Hmgr*
Arabidopsis thaliana	+	+	+	+	+	+	+	+
Andrographis paniculata	+	–	–	–	–	–	–	+
Antirrhinum majus	+	+	–	–	–	–	–	–
Artemisia annua	+	+	–	–	–	–	–	+
Catharanthus roseus	+	+	–	–	+	–	–	+
Camptotheca acuminate		+						+
Cistus incanus subsp. creticus		+						+
Chrysanthemum x morifolium	+	+	–	–	–	–	–	–
Citrus jambhiri						+		?
Croton stellatopilosus	+	+	–	–	–	–	–	–
Elaeis guineensis	+	+	–	–	–	–	–	–
Forsythia x intermedia	+	+	–	+	–	–	+	–
Ginkgo biloba	+	+	+	+	–	–	+	+
Hevea brasiliensis	+	+	–		–	+	–	+
Hordeum vulgare		+						–
Linum usitatissimum		+						–
Lycopersicon esculentum	+	+	–	+	+	–	–	+
Lycopersicon hirsutum	+	–	–	+	–	–	–	–
Mentha x piperita	+	+	–	+	–	–	–	–
Medicago truncatula	+	–	–	–	–	–	–	+
Mesostigma viride				+			+	?
Morinda citrifolia	+	–	–	–	–	–	–	–
Narcissus pseudonarcissus	+	–	–	–	–	–	–	–
Nicotiana benthamiana				+	+		+	–
Nicotiana tabacum		+						+
Oryza sativa (Japonica)	+	+	+	+	+	+	+	+
Oryza sativa (Indiaca)	+	+	+	–	–	+	–	+
Picrorhiza kurrooa		+		+				+
Plectranthus barbatus		+						–
Pueraria montana var. lobata	+	+	–	–	–	–	–	–
Rauvolfia verticillata		+		+				–
Salvia miltiorrhiza		+	+	+				+
Stevia rebaudiana	+	+	+	+	+	+	–	–
Tagetes erecta	+	+	–	–	–	–	–	+
Taxus chinensis		+						–
Taxus cuspidate		+						–
Taxus x media	+	+	–	–	–	+	–	+
Vitis Vinifera			+					+
Zea mays	+	+	–	–	+	–	–	+

dxs: 1-deoxy-D-xylulose 5-phosphate synthase; ***dxr***: 1-deoxy-D-xylulose 5-phosphate reductase; ***ispD***: 4-diphosphocytidyl2*C*-methyl-D-erythritol synthase; ***ispE***: 4-diphosphocytidyl 2*C*-methyl-D-erythritol kinase; **mecs**; 2*C*-methyl-D-erythritol 2,4-cyclodiphosphatesynthase; ***ispG***: 1-hydroxy-2-methyl-2-(*E*)-butenyl 4-diphosphate synthase; ***ispH***: 1-hydroxy-2-methyl-2-(*E*)-butenyl 4-diphosphate reductase; ***hmgr***: 3-hydroxy-3-methylglutaryl-CoA reductase.

Cragg, G.M., Newman, D.J., and Snader, K.M. (1997) Natural products in drug discovery and development. J. Nat. Prod. *60*, 52–60.

Croteau, R. (1987) Biosynthesis and catabolism of monoterpenoids. Chem. Rev. *87*, 929–954.

Dhingra, V., Rao, K.V., and Narasu, M.L. (2000) Current status of artemisinin and its derivatives as antimalaria drugs. Life Sci. *66*, 279–300.

Dixon. R.A. (2001) Natural products and plant disease resistance. Nature *411*, 843–847.

Dubey, V.S., Bhalla, R., and Luthra, R. (2003) An overview of the non-mevalonate pathway for terpenoid biosynthesis in plants. J. Biosciences *28*, 637–646.

Dudareva, N., Andersson, S., Orlova, I., Gatto, N., Reichel, M., Rhodes, D., Boland, W., and Gershenzon, J. (2005) The nonmevalonate pathway supports both monoterpene and sesquiterpene formation in snapdragon flowers. Proc. Natl. Acad. Sci. U.S.A. *102*, 933–938.

Eichinger, D., Bacher, A., Zenk, M.H., and Eisenreich, W. (1999) Analysis of metabolic pathways via quantitative prediction of isotope labeling patterns: a retrobiosynthetic ^{13}C NMR study on the monoterpene loganin. Phytochemistry *51*, 223–226.

Eisenreich, W., Menhard, B., Hylands, P.J., Zenk, M.H., and Bacher, A. (1996) Studies on the biosynthesis of taxol: The taxane carbon skeleton is not of mevalonoid origin. Proc. Natl. Acad. Sci. U.S.A. *93*, 6431–6436.

Eisenreich, W., Rohdich, F., and Bacher, A. (2001) Deoxyxylulose phosphate pathway to terpenoids. Trends Plant Sci. *6*, 78–84.

Eisenreich, W., Sagner, S., Zenk, M.H., and Bacher, A. (1997) Monoterpenoid essential oils are not of mevalonoid origin. Tetrahedron Lett. *38*, 3889–389.

Eisenreicha, W., Bachera, A., Arigonib, D., and Rohdicha, F. (2004) Biosynthesis of isoprenoids *via* the non-mevalonate pathway. Cell Mol. Life Sci. *61*, 1401–1426.

Engprasert, S., Shoyama, Y., and Taura, F. (2005) Molecular cloning, expression and characterization of recombinant 1-deoxy-D-xylulose-5-phosphate reductoisomerase from *Coleus forskohlii* Brig. Plant Sci. *169*, 287–294.

Fellermeier, M., Kis, K., Sagner, S., Maier, U., Bacher, A., and Zenk, M.H. (1999) Cell-free conversion of 1-deoxy-D-xylulose 5-phosphate and 2-C-methyl-D-erythritol 4-phosphate into β-carotene in higher plants and its inhibition by fosmidomycin. Tetrahedron Lett. *40*, 2743–2746.

Fowler, D.J., Hamilton, J.T.G., Humphrey, A.J., and O'Hagan, D. (1999) Plant Terpene Biosynthesis. The Biosynthesis of Linalyl Acetate in *Mentha citrate.* Tetrahedron Lett. *40*, 3803–3806.

Gao, S., Lin, J., Liu, X., Deng, Z., Li, Y., Sun, X., and Tang, K. (2006) Molecular Cloning, Characterization and Functional Analysis of a 2C-methyl-D-erythritol 2,4-cyclodiphosphate Synthase Gene from *Ginkgo biloba*. J. Biochem. Mol. Biol. *39*, 502–510.

Gong, Y., Liao, Z., Chen, M., Zuo, K., Guo, L., Tan, Q., Huang, Z., Kai, G., Sun,.X., Tan, F., and Tang, K. (2005) Molecular cloning and characterization of a 1-deoxy-D-xylulose 5-phosphate reductoisomerase gene from *Ginkgo biloba*. DNA Seq. *16*, 111–120.

Gong, Y.F., Liao, Z.H., Guo, B.H., Sun, X.F., and Tang, K.X. (2006) Molecular cloning and expression profile analysis of *Ginkgo biloba* DXS gene encoding 1-deoxy-D-xylulose 5-phosphate synthase, the first committed enzyme of the 2-C-methyl-D-erythritol 4-phosphate pathway. Planta Med. *72*, 329–335.

Guevara-Garcia, A., San Roman, C., Arroyo, A., Cortes, M.E., de la Luz Gutierrez-Nava, M., and Leon, P. (2005) Characterization of the Arabidopsis clb6 mutant illustrates the importance of posttranscriptional regulation of the methyl-D-erythritol 4-phosphate pathway. Plant Cell *17*, 628–643.

Han, Y.S., Roytrakul, S., Verberne, M.C., van der Heijden, R., Linthorst, H.J.M., and Verpoorte, R. (2003) Cloning of a cDNA encoding 1-deoxy-D-xylulose 5-phosphate synthase from *Morinda citrifolia* and analysis of its expression in relation to anthraquinone accumulation. Plant Sci. *164*, 911–917.

Hans, J., Hause, B., Strack, D., and Walter, M.H. (2004) Cloning, characterization, and immunolocalization of a mycorrhiza-inducible 1-deoxy-D-xylulose 5-phosphate reductoisomerase in arbuscule-containing cells of maize. Plant Physiol.134, 614–624.

Harborne, J.B. (2001) Twenty-five years of chemical ecology. Nat. Prod. Rep. *18*, 361–379.

Hertewich, U., Zapp, J., Becker, H., and Adam, K.P. (2001) Biosynthesis of a hopane triterpene and three diterpenes in the liverwort *Fossombronia alaskana*. Phytochemistry *7*, 1049–1054.

Hirai, N., Yoshida, R., Todoroki, Y., and Ohigashi, H. (2000) Biosynthesis of abscisic acid by the non-mevalonate pathway in plants, and by the mevalonate pathway in fungi. Biosci. Biotechnol. Biochem. *64*, 1448–1458.

Jeffrey, F.M., Rajagopal, A., Malloy, C.R., and Sherry, A.D. (1991) ^{13}C-NMR: a simple yet comprehensive method for analysis of intermediary metabolism. Trends Biochem. Sci. *6*, 5–10.

Jennewein, S., Wildung, M.R., Chau, M., Walker, K., and Croteau, R. (2004) Random sequencing of an induced Taxus cell cDNA library for identification of clones involved in Taxol biosynthesis. Proc. Natl. Acad. Sci. U.S.A. *101*, 9149–9154.

Jin, H., Gong, Y., Guo, B., Qiu, C., Liu, D., Miao, Z., Sun, X., and Tang, K. (2006) Isolation and characterization of a 2C-methyl-D-erythritol 2,4-cyclodiphosphate synthase gene from *Taxus media*. Mol. Biol. (N.Y.) *40*, 914–921.

Kelleher, J.K. (2001) Flux estimation using isotopic tracers: common ground for metabolic physiology and metabolic engineering. Metab. Eng. *3*, 100–110.

Khemvong, S., and Suvachittanont, W. (2005) Molecular cloning and expression of a cDNA encoding 1-deoxy-D-xylulose-5-phosphate synthase from oil palm in Elaeis guineensis Jacq. Plant Sci. *169*, 571–578.

Kim, S.M., Kuzuyama, T., Chang, Y.J., Kwon, H.J., and Kim, S.U. (2006) Cloning and functional characterization of 2-C-methyl-D-erythritol 4-phosphate cytidyltransferase (GbMECT) gene from *Ginkgo biloba*. Phytochemistry *67*, 1435–1441.

Kishimoto, S., and Ohmiya, A. (2006) Regulation of carotenoid biosynthesis in petals and leaves of chrysanthemum (*Chrysanthemum morifolium*) Physiol. Plantarum. *128*, 436–447.

Klink, J.W., Becker, H., and Perry, N.B. (2005) Biosynthesis of irregular diterpenes in *Anisotome layallii* by head to tail coupling of geranyl diphosphate. Org. Biol. Chem. *3*, 542–545.

Knoss, W., Reuter, B., and Zapp, J. (1997) Biosynthesis of the labdane diterpenoid marrubin in *Marrubium vulgare* via a nonmevalonate pathway. Biochem. J. *326*, 449–454.

Lange, B.M., and Croteau, R. (1999) Isoprenoid biosynthesis via a mevalonate-independent pathway in plants: cloning and heterologous expression of 1-deoxy-D-xylulose-5-phosphate reductoisomerase from peppermint. Arch. Biochem. Biophys. *365*, 170–174.

Lange, B.M., and Croteau, R. (1999) Isopentenyl diphosphate biosynthesis via a mevalonate-independent pathway: isopentenyl monophosphate kinase catalyzes the terminal enzymatic step. Proc. Natl. Acad. Sci. U.S.A. *96*, 13714–13719.

Leite, A.C., Lopes, A.A., Kato, M.J., Bolzania, V.S., and Furlan, M., (2007) Biosynthetic Origin of the Isoprene Units in Chromenes of *Piper aduncum* (Piperaceae) J. Braz. Chem. Soc. *18*, 1500–1503.

Liao, Z., Chen, R., Chen, M., Yang, C., Wang, Q., and Gong, Y. (2007) A new 1-deoxy-D-xylulose 5-phosphate reductoisomerase gene encoding the committed-step enzyme in the MEP pathway from *Rauvolfia verticillata.* Z. Naturforsch C. J. Biosci. *62*, 296–304.

Lichtenthaler, H.K. (2001) Non-mevalonate isoprenoid biosynthesis: enzymes, genes and inhibitors. Biochem. Soc. Transact. *28*, 785–789.

Lichtenthaler, H.K., Schwender, J., Disch, A., and Rohmer, M. (1997) Biosynthesis of isoprenoids in higher plant chloroplasts proceeds *via* a mevalonate-independent pathway. FEBS Lett. *400*, 271–274.

Lichtenthaler, H.K. (1999) The 1-deoxy-D-xylulose-5-phosphate pathway of isoprenoid biosynthesis in plants. Ann. Rev. Plant Physiol. Plant Mol. Biol. *50*, 47–65.

Lichtenthaler, H.K., Zeidler, J., Schwender, J., and Muller, C. (2000) The non-mevalonate isoprenoid biosynthesis of plants as a test system for new herbicides and drugs against pathogenic bacteria and the malaria parasite. Z. Naturforsch. *55*, 305–313.

Lin, X., and Kaul, S. (1999) Sequence and analysis of chromosome 2 of the plant *Arabidopsis thaliana*. Nature *402*, 761–768.

Lois, L.M., Rodriguez-Concepcion, M., Gallego, F., Campos, N., and Boronat, A. (2000) Carotenoid biosynthesis during tomato fruit development: regulatory role of 1-deoxy-D-xylulose 5-phosphate synthase. Plant J. *22*, 503–513.

Luan, F., and Wust, M. (2002) Differential incorporation of 1-deoxy-D-xylulose into (3S)-linalool and geraniol in grape berry exocarp and mesocarp. Phytochemistry *60*, 451–459.

Maier, W., Schneider, B., and Strack, D. (1998) Biosynthesis of sesquiterpenoid cyclohexane derivatives in mycorrhizal barley roots proceeds via the glyceraldehydes-3-phosphate/pyruvate pathway. Tetrahedron Lett. *39*, 521–524.

Milborrow, B.V., and Lee, H.S. (1998) Endogenous biosynthetic precursors of (+) abscisic acid. VI. Carotenoids and abscisic acid formed by the non-mevalonate triose-pyruvate pathway in chloroplasts. Aust. J. Plant. Physiol. *25*, 507–512.

Moehs, C.P., Tian, L., Osteryoung, K.W., and Dellapenna, D. (2001) Analysis of carotenoid biosynthetic gene expression during marigold petal development. Plant Mol. Biol. 45, 281–293.

Ohyanagi, H., Tanaka, T., Sakai, H., Shigemoto, Y., Yamaguchi, K., Habara, T., Fujii, Y., Antonio, B.A., Nagamura, Y., Imanishi, T., Ikeo, K., Itoh, T., Gojobori, T., and Sasaki, T. (2006) The Rice Annotation Project Database (RAP-DB): hub for Oryza sativa ssp. japonica genome information. Nucleic Acids Res. *34*, D741-D744.

Page, J.E., Hause, G., Raschke, M., Gao, W., Schmidt, J., Zenk, M.H., and Kutchan, T.M. (2004) Functional Analysis of the Final Steps of the 1-Deoxy-D-xylulose 5-phosphate (DXP) Pathway to Isoprenoids in Plants Using Virus-Induced Gene Silencing. Plant Physiol. *134*, 1401–1413.

Phillips, M.A., Walter, M.H., Ralph, S.G., Dabrowska, P., Luck, K., Uros, E.M., Boland, W., Strack, D., Rodriguez-Concepcion, M., Bohlmann, J., and Gershenzon, J. (2007) Functional identification and differential expression of 1-deoxy-D-xylulose 5-phosphate synthase in induced terpenoid resin formation of Norway spruce (*Picea abies*) Plant Mol. Biol. *65*, 243–257.

Rieder, C.H., Jaun, B., and Arigoni, D. (2000) On the early steps of cineol biosynthesis in *Eucalyptus globules.* Helv. Chim. Acta. 83, 2504–2513.

Rodriguez-Concepcion, M., Ahumada, I., Diez-Juez, E., Sauret-Gueto, S., Lois, L.M., Gallego, F., Carretero-Paulet, L., Campos, N., and Boronat, A. (2001) 1-Deoxy-D-xylulose 5-phosphate reductoisomerase and plastid isoprenoid biosynthesis during tomato fruit ripening. Plant J. *27*, 213–222.

Rodrıguez-Concepcion, M., and Boronat, A. (2002) Elucidation of the Methylerythritol Phosphate Pathway for Isoprenoid Biosynthesis in Bacteria and Plastids. A Metabolic Milestone Achieved through Genomics. Plant Physiol. *130*, 1079–1089.

Rohdich, F., Wungsintaweekul, J., Luttgen, H., Fischer, M., Eisenreich, W., Schuhr, C.A., Fellermeier, M., Schramek, N., Zenk, M.H., and Bacher, A. (2000) Biosynthesis of terpenoids: 4-diphosphocytidyl-2-C-methyl-D-erythritol kinase from tomato. Proc. Natl. Acad. Sci. U.S.A. *97*, 8251–8256.

Rohmer, M. (2003) Mevalonate-independent methylerythritol phosphate pathway for isoprenoid biosynthesis. Elucidation and distribution. Pure Appl. Chem. *75*, 375–387.

Rohmer, M. (1999) The discovery of a mevalonate independent pathway for isoprenoid biosynthesis in bacteria, algae and higher plants. Nat. Prod. Rep. *16*, 565–574.

Sagner, S., Eisenreich, W., Fellermeier, M., Latzel, C., Bacher, A., and Zenk, M.H. (1998) Biosynthesis of 2-C-methyl-D-erythritol in plants by rearrangement of the terpenoid precursor, 1-deoxy-D-xylulose-5-phosphate. Tetrahedron Lett. *39*, 2091–2094.

Sasaki, T., and Matsumoto, T. (2002) The genome sequence and structure of rice chromosome 1. Nature *420*, 312–316.

Sato, S., Nakamura, Y., Kaneko, T., Katoh, T., Asamizu, E., and Tabata, S. (2000) Structural analysis of *Arabidopsis thaliana* chromosome 3. I. Sequence features of the regions of *4*, *504*, 864 bp covered by sixty P1 and TAC clones. DNA Res. *7*, 131–135.

Schwender, J., Muller, C., Zeidler, J., and Lichtenthaler, H.K. (1999) Cloning and heterologous expression of a cDNA encoding 1-deoxy-D-xylulose-5-phosphate reductoisomerase of *Arabidopsis thaliana*. FEBS Lett. *455*, 140–144.

Scolink, P.A. and Bartley, G.E. (1996) A table of cloned genes involved in Isoprenoid biosynthesis. Plant Mol. Biol. Report *14*, 305–319.

Seetang-Nun, Y., Sharkey, T.D., and Suvachittanont, W. (2008) Isolation and characterization of two distinct classes of DXS genes in *Hevea brasiliensis*. DNA Seq. *19*, 291–300.

Seetang-Nun, Y., Sharkey, T.D., and Suvachittanont, W. (2008) Molecular cloning and characterization of two cDNAs encoding 1-deoxy-D-xylulose 5-phosphate reductoisomerase from *Hevea brasiliensis*. J. Plant Physiol. *165*, 991–1002.

Seki, M., Iida, K., Satou, M., Sakurai, T., Akiyama, K., Ishida, J., Nakajima, M., Enju, A., Kamiya, A., Narusaka, M., Carninci, P., Kawai,J., Hayashizaki,Y., and Shinozaki, K. (2002). *Arabidopsis thaliana* full-length cDNA. Published Only in Database. http://www.ncbi.nlm.nih.gov/entrez/viewer.fcgi.

Sharkey, T.D., Yeh, S., Wiberley, A.E., Falbel, T.G., Gong, D., and Fernandez, D.E. (2005) Evolution of the Isoprene Biosynthetic Pathway in Kudzu. Plant Physiol. *137*, 700–712.

Sprenger G.A. (1996) Carbohydrate metabolism in *Zymomonas mobilis*: a catabolic highway with some scenic routes. FEMS Microbiol. Lett. *145*, 301–307.

Steliopoulos, P., Wust, M., Adam, K.P., and Mosandl, A. (2002) Biosynthesis of the sesquiterpene germacrene D in *Solidago canadensis*: 13C and 2H lebelling studies. Phytochemistry *60*, 13–20.

Thiel, R., and Adam, K.P. (2002) Incorporation of [1–13C]1-deoxy-D-xylulose into isoprenoids of the liverwort *Conocephalum conicum*. Phytochemistry *59*, 269–274.

Totte, N.M.L.C., Van den Ende, W., Van Damme, E.J.M., Compernolle, F., Baboeuf, I., and Geuns, J.M.C. (2003) Cloning and heterologous expression of early genes in gibberellin and steviol biosynthesis via the methylerythritol phosphate pathway in *Stevia rebaudiana*. Plant Physiol. *81*, 517–522.

Veau, B., Courtois, M., Oudin, A., Chenieux, J.C., Rideau, M., and Clastre, M. (2000) Cloning and expression of cDNAs encoding two enzymes of the MEP pathway in *Catharanthus roseus*. Biochem. Bioph. Acta. *1517*, 159–163.

Verpoorte, R., and Memelink, J. (2002) Engineering secondary metabolite production in plants. Curr. Opin. Biotechnol. *13*, 181–187.

Wu, S.J., Shi, M., and Wu, J.Y. (2009) Cloning and characterization of 1-deoxy-D-xylulose 5-phosphate reductoisomerase gene for diterpenoid tanshinone biosynthesis in *Salvia miltiorrhiza* hairy roots. Biotechnol. Appl. Biochem. *52*, 89–95.

Wungsintaweekul, J., Sirisuntipong, T., Kongduang, D., Losuphanporn, T., Ounaroon, A., Tansakul, P., and De-Eknamkul, W. (2008) Transcription profiles analysis of genes encoding 1-deoxy-D-xylulose 5-phosphate synthase and 2C-methyl-D-erythritol 4-phosphate synthase in plaunotol biosynthesis from *Croton stellatopilosus*. Biol. Pharm. Bull. *31*, 852–856.

Yao, H., Gong, Y., Zuo, K., Ling, H., Qiu, C., Zhang, F., Wang, Y., Pi, Y., Liu, X., Sun, X., and Tang, K. (2008) Molecular cloning, expression profiling and functional analysis of a DXR gene encoding 1-deoxy-D-xylulose 5 phosphate reductoisomerase from *Camptotheca acuminate*. J. Plant Physiol. *165*, 203–213.

Yu, J., Wang, J., Lin, W., Li, S., and Li, H. (2005) The Genomes of *Oryza sativa*: A History of Duplications. PLoS. Biol. *3*, E38-E42.

Zeidler, J.G., Lichtenthaler, H.K., May, H.U., and Lichtenthaler, F.W. (1997) Is isoprene emitted by plants synthesized via a novel isopentenyl pyrophosphate pathway? Z. Naturforsch. *52C*, 15–23.

Zeidler, J.G., Schwender, J., Muller, C., Wiesner, J., Weidemeyer, C., Beck, E., Jomaa, H., and Lichtenthaler, H.K. (1998) Inhibition of the non-mevalonate 1-deoxy-D-xylulose-5-phosphate pathway of plant isoprenoid biosynthesis by fosmidomycin. Z. Naturforsch. *53C*, 980–986.

Chapter 6

Establishment of the Regeneration System for *Vicia faba* L.

Shimaa Bahgat, Omer A. Shabban, Osama El-Shihy, David A. Lightfoot, and Hany A. El- Shemy

Abstract

A reliable regeneration system for faba bean has been difficult to establish and therefore, the genetic improvement of *Vicia faba* L. was delayed. The paper describes a method of somatic embryo induction in callus of *V. faba*. Two Egyptian faba bean cultivars 'Giza 2' and '24 Hyto' were used. Callus was induced from epicotyls and shoot tips cultured on MS or Gamborg medium supplemented with 3% sucrose and 0.025% (w/v) for each of ascorbic and citric acid, 0.8% agar and different concentrations of 10 mg/l BAP, 0.5 mg/l of each NAA and 2,4-dichlorophenoxyacetic acid (M1) and 1 mg/l BAP and 0.5 mg/l NAA (M2) . The media with BAP, NAA and 2,4-D were optimal for embryogenic callus induction. Somatic embryos developed after transfer of the callus to 1/2 B5 medium with no plant growth regulators. There were various stages of somatic embryo development present including globular, heart-shaped, torpedo, and cotyledonary stages. Embryos developed into plantlets and plants were regenerated. RAPD analyses were performed to investigate the genetic stability of the regenerated plants obtained from different treatments and different explants. The cultivar Giza 2 exhibited more genetic stability than cultivar 24 Hyto. In conclusion, a regeneration system was established suitable for both gene transformation and the isolation of somaclonal mutants. The regeneration system will be used in order to improve the nutritional value of faba bean.

Introduction

Legumes are the third largest family of dicotyledons, *Leguminosae*. Examples include soybean (*Glycine max)*, pea (*Pisum sativum*), faba bean (*Vicia faba*), mung bean (*Vigna radiata*), chickpea (*Cicer arietinum*) and common bean (*Phaseolus vulgaris*). Faba bean (*Vicia faba* L.) is one of the most important grain Legumes in the world. Because Faba bean is an excellent source of protein, it is an important pulse crop used for human consumption and animal feeding, faba bean also plays a great role in the biological fixation of aerial nitrogen (Duke, 1981; Jelenić *et al.*, 2000). This species alone occupies nearly 3.2×10^6 ha worldwide in 1991 (FAO statistics, 1992) with a world production close to 4.5 million tons in 2004 (Gutierrez, 2006). Even though faba bean has been cultivated in many countries, 49% of the total production comes from China followed by Egypt (10, 8%), Ethiopia (8.5%), and Australia by 5.8% in 2000 (FAO, statistics 2001). Unfortunately, its susceptibility to environmental conditions, biotic and abiotic stress and instability of the yield made this crop less attractive. Also, difficulties in pollination control and the limited genetic pool has led to slow down progress in varietal improvement (Bond, 1987; Selva *et al.*, 1989). Today to improve faba against biotic and abiotic stress and to improve its nutritional value with sulphur amino acids, traditional breeding programs could be supplemented by gene transfer technology which requires an efficient and suitable regeneration system, which occurs via organogenesis or embryogenesis, either directly from the excised tissue or indirectly after formation of callus depending on internal and external plant factors (Kuchuk, 2001). In tissue culture faba bean appears to be recalcitrant towards *in vitro* regeneration (Khalafalla and Hattori, 2000), due to difficulties in regeneration from callus, and the releasing of phenolic compounds, so studies on *in vitro* culture of faba bean are therefore difficult to carry out (Böttinger *et al.*, 2001). Regeneration system of faba bean through indirect somatic embryogenesis was reported by Griga *et al.* (1987) they were used the cotyledons on media supplemented with 2,4D. Somatic embryos at globular stage to early torpedo stage were released, but their development was terminated at the late-torpedo stage. In tissue culture, the regenerated plants from cultured cells are known to exhibit genetic and epigenetic changes collectively called somaclonal variations (Razdan, 1994). In this study we described a new method for regeneration of faba bean through somatic embryogenesis, with a development to complete plants. RAPD-PCR molecular markers were performed to identify the genetic variability between the control plants and the regenerated plants of each cultivar.

Materials and methods

Plant material

Two cultivars of faba bean Giza 2 and 24 Hyto (cv. Giza 2 was obtained from Agricultural Research Center, Egypt and cv. 24 Hyto was obtained from Agronomy Department,

Faculty of Agriculture, Cairo University, Egypt) were used for this study. Segments of shoot tips and epicotyls were the sources of *in vitro* experimental materials.

Preparation of explants

Seeds of faba bean were sterilized in 70% (v/v) ethanol for 30 second, followed by 20% (v/v) commercial bleach for 15 min and washed 3–4 times with sterile distilled water. Sterilized seeds were germinated on half-strength MS medium (Murashige and Skoog, 1962) supplemented with 15 g/l sucrose and solidified with 8 g/l agar. The seeds were germinated in darkness at 25°C. After a week shoot tips and epicotyls were cut with scalpel blade into segments of 5–7 mm before cultivation on callus induction media.

Culture media

Two different treatments (M1& M2) were used in this study. The MS medium and Gamborg medium (Gamborg *et al.*1968) were used for plant tissue culture. All of callus and regeneration media were supplemented with 3% (w/v) sucrose and 0.025% (w/v) for each of ascorbic and citric acid (Sigma, USA). The pH of all cultures was adjusted to 5.7 then solidified with 0.8% (w/v) agar and autoclaved at 121°C for 25 min. The cultures were incubated at 25°C ± 1 under a 16/8 h light/dark photoperiod regime provided by white fluorescent light at 716 Lux.

M1 treatment

Epicotyls and shoot tips of the two cultivars were placed in callus induction medium CIM1 which contained MS medium with 10 mg/l BAP, 0.5 mg/l of each NAA and 2, 4D (Sigma, USA), and sub-cultured every 3–4 weeks, after two months calluses were transferred to a regeneration medium RM1 containing Gamborg medium or MS medium supplemented with B5 vitamins and 2 mg/l BAP. When globular structures appeared on the surface of callus with a white-greenish brilliant color, or a single embryo with a long structure had formed, they were separated from the callus and transferred to RM1 medium supplemented with 20 mg/l of glutamine (Sigma, St Louis, MO, USA). After 4–6 weeks, Nemours somatic embryos were formed (at globular stages) on an embryogenic callus, they were transferred to hormone free medium 1/2 B5 to encourage the development of the somatic embryos into plants. They were sub-cultured every 4 weeks. The hormone-free medium caused the formation of hundreds of somatic embryos. Each was separated and allowed to develop into a plant. Normally they formed shoots and roots in the same medium. Such normal plantlets were transferred to the soil, covered at first for adaptation to the glasshouse. Covers were gradually removed to allow hardening. Plants were allowed to flower and self fertilize.

M2 treatment

Epicotyls and shoot tips of the faba bean cvs. were placed in the callus induction medium CIM2 which contained MS medium with 1 mg/l BAP and 0.5 mg/l NAA. Within 6–7 weeks calluses were formed and embryo-like structures started to be formed. The embryo-like structures were excised and transferred to the regeneration medium RM2, containing MS medium supplemented with B5 vitamins, 1.2 mg/l BAP, and 0. 5 mg/l NAA. Somatic embryos were formed at different stages. They were excised when they appeared to be at the beginning of the cotyledonary stage. Embryos developed to complete plants in 1/2 B5 medium without plant growth regulators. After the elongation of shoots and roots, plantlets were transferred to the soil to prepare them for adaptation to glass-house conditions.

RAPD analysis

Total DNA was isolated from each control and regenerated plant using the CTAB protocol (Doyle and Doyle, 1987). PCR reactions were performed in total volume of 25µl and 6 random Primers (from the OPA set) were used as described in Table 1 (all primers were obtained from Operon Technologies Inc., Alameda, CA, USA). DNA amplification was carried out in a T-gradient thermal cycler (Biometra, Göttingen, FRG), PCR reactions were hot –started at 95°C for 4 min and subjected to 40 cycles as follows; 1 min at 94°C; 2 min at 35°C; and 2 min at 72°C. The last extension phase was prolonged to 7 min at 72°C. PCR products were separated by electrophoresis in 1.5 % agarose gel and visualized by Ultraviolet illumination after staining with ethidium bromide, and photographed using gel-documentation system (BIO-RAD, Hercules CA).

Results

Callus tissues were initiated from epicotyls segments and shoots tip of two Egyptian faba bean cvs. Giza 2 and 24 Hyto using two different treatments M1 and M2. All the induction media were supplemented with various concentrations of auxins and cytokinins, segments of shoot tips and epicotyls induced a mixture of greenish, brownish and yellowish compact callus (Fig. 1A-F).

The M1 treatments consisted of CIM1 (callus induction medium) and RM1 (regeneration medium). In CIM1 shoot tips and epicotyls of each cultivar were cultured for two months during which time they were sub-cultured every 3–4 weeks. They had produced a compact yellowish callus; after two months the calluses were transferred to regeneration medium RM1. After 3–7 weeks, structures that looked like an embryo with long structure were obtained, or sometimes a globular smooth structures with white-greenish transparent color, were formed on the surface of callus-derived from shoot tips. Cultivar Giza 2 formed these structures after 3 weeks while cv. 24 Hyto formed the same structures after 7 weeks of culturing in RM1 medium. The embryo-like structures were separated and cultured in the same RM1 medium, but supplemented with 20 mg/l glutamine. Within a week these embryo

Table 1. Nucleotide sequences of six primers used in RAPD analysis.

Sequence (5′-3′)	Primer
AATCGGGCTG	OPA04
AGGGGTCTTG	OPA05
GAAACGGGTG	OPA07
GGGTAACGCC	OPA09
GTGATCGCAG	OPA10
CAGCACCCAC	OPA13

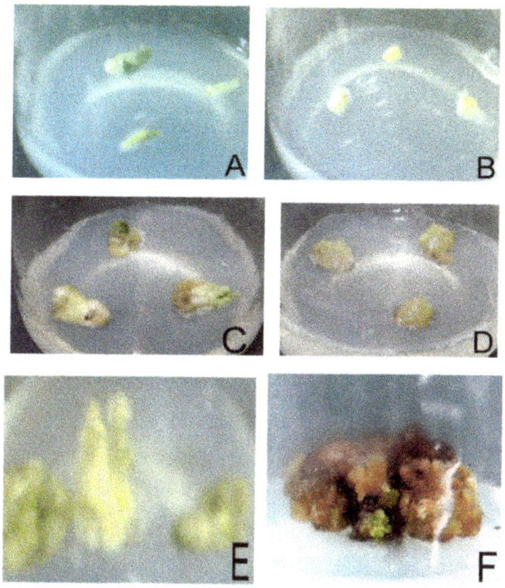

Fig. 1 Examples of embryogenic callus from *V. faba* cvs. Giza 2 and 24 Hyto. Panels **A** and **B** show segments of shoot tips and epicotyls. Panels **C** and **D** show segments of shoot tips and epicotyls on callus induction media after 10–15 days. Panels **E** and **F** show shoot tips and epicotyls respectively after two months on callus induction medium. Panel **E** shows callus derived from shoot tips of cv. 24 Hyto on CIM1 medium. Panel **F** shows callus derived from an epicotyl of cv. 24 Hyto in CIM2 medium which turned brown due to phenolic compounds.

structures grew and developed into globular stage (Fig. 2F, G), 3 weeks later, a lot of somatic embryos with white-greenish color were formed in clusters with embryogenic callus (Fig. 2H). For further development these embryos were transferred to hormone free medium (1/2 B5 medium). It was also noted that, the hormone free medium enhanced the production of new generations of somatic embryos. Each embryo was capable of producing a lot of embryos over a period of 14 months. Therefore there was asynchronous development of embryos to complete plants over that time. The somatic embryos which derived from shoot tips of cv. Giza 2 readily developed to fertile plants, while the cultivar 24 Hyto did not, being arrested at the torpedo stage. Although the embryos of cv. Giza 2 developed slowly, this cultivar produced more embryos than that of cv. 24 Hyto (Table 2).

Epicotyls from cv. Giza 2 which cultured on M1 also formed embryogenic callus with somatic embryos in the globular stage. However, these globular structures did not develop further and eventually became contaminated. The experiment was repeated many times without any success. Epicotyls from cv. 24 Hyto formed callus but not embryos.

The other treatment (M2) induced callus from epicotyls of each cv. but failed to induce callus from shoot tips where, shoots formed directly (Fig. 3A). These shoots were transferred to 1/2 B5 medium supplemented with 3 mg/l IBA for rooting. Epicotyls from cv. Giza 2 induced small calluses with a greenish- yellow color, but all failed to produce any somatic embryos. Only a few adventitious shoots were obtained from buds on the callus surface (Fig. 3B).

Calluses derived from epicotyls of 24 Hyto formed creamy- brownish masses (Fig. 1F), that increased in weight rapidly (Table 5). These calluses were embryogenic after 7 weeks of culturing on CIM2 medium (Fig. 2A). To develop the embryos, calluses were transferred to RM2 medium (Fig. 2B). Calluses developed and produced somatic embryos in clusters at different growth stages including globular and heart stages and the early torpedo stages (Fig. 2C). The embryos were separated at the beginning of the cotyledonary stage (Fig. 2D), and each transferred to hormone free medium (1/2 B5) for their development. The hormone free medium increased the numbers of embryos formed by stimulating many generations of somatic embryos with different stages forming on the first embryo explants. The adventitious embryos developed in the same medium to become young plantlets (Fig. 2E). These plantlets then transferred to the green house for adaptation.

The number of total shoots which developed from somatic embryos varied between each cultivar. In cultivar Giza 2, the calluses that derived from shoot tips were reproducible and produced a lot of somatic embryos. Hence within 8 months after the formation of the first embryo, 400 shoots per callus were developed from somatic embryos. However, in cultivar 24 Hyto, within 8 months, only 127 shoots had developed from somatic embryos derived from the epicotyl calluses.

Therefore cultivar differences and explant differences play important roles determining both callus formation and regeneration percentages; the type of structures that develop callus and the regeneration frequency of those structures (Tables 2 and 3).

A significant response towards callus, and regeneration percentage were observed, due to the hormonal compositions of each treatment (Table 4). Although the M2 medium had the highest value of the

Table 2. Effect of cultivar generally on fresh weight of callus, Callus (%) and regeneration through somatic embryogenesis (%).

Cultivar	Fresh weight of callus	Callus (%)	Calluses producing somatic embryos (%)
Giza 2	2.14[b]	69.38[b]	4.14[a]
24 Hyto	3.69[a]	72.91[a]	3.31[b]

Means followed by the same letter are not significantly different at *p*≤*0.05* using Duncan's multiple range test.

Table 3. Effect of explants generally on, fresh weight of callus, callus (%) and regeneration through somatic embryogenesis (%) for both cultivars (Giza 2 and 24 Hyto).

Type of explant	Fresh weight of callus	Callus (%)	Calluses producing somatic embryos (%)
Shoot tip	1.64[b]	49.58[b]	6.76[a]
Epicotyl	4.19[a]	92.71[a]	0.70[b]

Means followed by the same letter are not significantly different at *p*≤0.05 using Duncan's multiple range test.

Fig. 2. *In vitro* regeneration through indirect somatic embryogenesis from *V.faba* cvs. Giza 2 and 24Hyto. **A–E** Somatic embryos derived from calluses of epicotyls of cv. 24 Hyto. **A** Embryogenic cells formed on callus derived from epicotyls in CIM2 medium. **B** Differentiation of the embryogenic cells in RM2 medium. **C** Different stages of somatic embryos were formed in RM2 medium including globular, heart and torpedo stages. **D** Maturation of somatic embryos into the cotyledonary stage in hormone free medium with asynchronous development. **E** The development of somatic embryos to young plantlets. **F–N** Different stages of somatic embryos of cv. Giza 2 which formed on RM1 medium from calluses derived from shoot tips. **F–G** The stages of embryos formation including formation of globular stage from a single embryo, this process took place in RM1 medium supplemented with 20 mg/l glutamine. **H** White-green structures identified as the early stages of somatic embryos were obtained on the previous medium including formation of embryogenic callus, these embryos were transferred subsequently to hormone free medium for their development. **I–K** Different stages of somatic embryos were formed in clusters in hormone free medium. **L** A cluster of somatic embryos developed before separating. **M** Development of embryos to young plantlets with shoots and roots in the hormone free medium. **N** Rooted plants that were transferred to the soil for acclimatization.

Fig. 3. **A** Direct organogenesis was formed from the shoot tips of cv. Giza 2 on CIM2 medium. **B** Adventitious shoots were formed from callus of epicotyls of cv. Giza 2 on M2 treatment.

callus fresh weight, the M1 treatment was the most effective in each of regeneration and callus percentages, so we detected that there was not any relationship between the weight of callus and the frequency of the regeneration.

Also the interactions between the three elements of this study (cultivars, explants and treatments) in all of the callus and regeneration percentages, showed that the best explant which led to the highest percentage of regeneration was the shoot tip of cv. Giza 2 using M1 treatment (Tables 5–8).

RAPD analysis

RAPD analyses were performed to investigate the genetic stability of the regenerated plants of each cultivar obtained from different treatments and different explants. Therefore six random primers were used (Table 1) to amplify DNA from many genomic locations. The 100 bp DNA molecular marker (with range of 100, 200, 300, 400, 500, 600, 700, 800, 900 and 1,000 bp) was used in this experiment. Among the RAPD profiles (Fig. 4) the cultivar Giza 2 showed identity between the control and the earliest regenerated plant in all profiles, except for new fragments detected at 1000 bp (OPA-04), 550 bp (OPA-05), and 500 bp of (OPA-07). Cultivar 24 Hyto showed identity in DNA banding patterns between the control and the earliest regenerated plant only with primer OPA-13. The other 5 primers derived DNA banding patterns showed new fragments at 300 bp (OPA-05) and 400 bp (OPA-09), whereas some fragments were absent at 750 bp (OPA-04), 450 bp (OPA-07) and 800 bp (OPA-10). Therefore, the cultivar Giza 2 exhibited more genetic stability than cultivar 24 Hyto.

Table 4. Effect of treatments generally on fresh weigh of callus, callus (%) and regeneration through somatic embryogenesis(%) for both cultivars (Giza 2 and 24 Hyto).

Treatment	Fresh weight of callus	Callus (%)	Calluses producing somatic embryos (%)
M1	2.66^a	97.52^a	6.89^a
M2	3.17^a	44.78^b	0.56^b

Means followed by the same letter are not significantly different at $p \leq 0.05$ using Duncan's multiple range test.

Table 5. Effect of interaction between treatments and cultivars on fresh weight of callus, callus (%), and regeneration through somatic embryogenesis (%).

Treatment	Cultivar	Fresh weight of callus	Callus (%)	Calluses producing somatic embryos (%)
M1	Giza 2	2.99^a	96.20^b	8.28^a
	24 Hyto	2.33^a	98.83^a	5.50^b
M2	Giza 2	1.29^a	42.57^d	0.00^d
	24 Hyto	5.05^a	46.98^c	1.12^c

Means followed by the same letter are not significantly different at $p \leq 0.05$ using Duncan's multiple range test.

Table 6. Effect of interaction between treatments and explants on fresh weight of callus, callus (%), and regeneration through somatic embryogenesis (%) for both cultivars (Giza 2 and 24 Hyto).

Treatment	Explant	Fresh weight of callus	Callus (%)	Calluses producing somatic embryos (%)
M1	Shoot tip	3.28^b	99.17^a	13.51^a
	Epicotyl	2.04^c	95.87^b	0.27^c
M2	Shoot tip	0.00^d	0.00^d	0.00^c
	Epicotyl	6.35^a	89.55^c	1.12^b

Means followed by the same letter are not significantly different at $p \leq 0.05$ using Duncan's multiple range test.

Table 7. Effect of interaction between cultivars and explants on fresh weight of callus, callus (%), and regeneration through somatic embryogenesis (%).

Cultivar	Explant	Fresh weight of callus	Callus (%)	Calluses producing somatic embryos (%)
Giza 2	Shoot tip	1.71^c	49.33^c	8.01^a
	Epicotyl	2.57^b	89.43^b	0.27^d
24 Hyto	Shoot tip	1.57^c	49.83^c	5.50^b
	Epicotyl	5.82^a	95.98^a	1.12^c

Means followed by the same letter are not significantly different at $p \leq 0.05$ using Duncan's multiple range test.

Discussion

This study describes a new reproducible system for plant regeneration via somatic embryogenesis using callus cultures derived from shoot tips and epicotyls of two Egyptian cultivars of *V. faba*. Many investigators have suggested that the difficulties of indirect regeneration from faba bean were due to the accumulation of phenolic compounds, which led to cell death (Böttinger *et al.*, 2001), Here it was shown to be very important to avoid the accumulation of the phenolic compounds within the first two weeks of culturing the explants on the callus induction media. However some calluses that turned to brown or black colors later (Fig. 1 F), presumably due to the phenolic compounds, still showed a good ability to divide and produce several generations of somatic embryos. (Fig. 2 A, B, I, G). The addition of ascorbic and citric acids to the media helped to decrease or prevent the production or oxidation of the phenolic compound, therefore enhancing the production of somatic embryos. The addition of citric acid to media was reported by Anthony *et al.*, 2004, to

Table 8. Effect of interactions among treatments, cultivars and explants on fresh weight of callus, callus (%), and regeneration through somatic embryogenesis (%).

Treatment	Cultivar	Explant	Fresh weight of callus	Callus (%)	Calluses producing somatic embryos (%)
M1	Giza 2	Shoot tip	3.42^{b}	98.67^{a}	16.02^{a}
		Epicotyl	2.55^{b}	93.73^{b}	0.53^{d}
	24 Hyto	Shoot tip	3.13^{b}	99.67^{a}	11.00^{b}
		Epicotyl	1.53^{c}	98.00^{a}	0.00^{d}
M2	Giza 2	Shoot tip	0.00^{d}	0.00^{d}	0.00^{d}
		Epicotyl	2.59^{b}	85.13^{c}	0.00^{d}
	24 Hyto	Shoot tip	0.00^{d}	0.00^{d}	0.00^{d}
		Epicotyl	10.11^{a}	93.97^{b}	2.25^{c}

Means followed by the same letter are not significantly different at $p \leq 0.05$ using Duncan's multiple range test.

prevent the browning and to enhance the formation of somatic embryos in *Conostephium pendulum*. Also the addition of ascorbic acid was beneficial for long-term common bean (*Phaseolus vulgaris*) callus maintenance (Mohamed *et al.*, 1993).

In this study, callus induction media used enrichment with cytokinin (as BAP) that was higher in concentration than auxins in each treatment. Cytokinin was important for the formation of callus and initiation of somatic embryogenesis. Recently, attempts to regenerate plants via somatic embryogenesis focused on pre-cultivation of the initial plants on cytokinin- enriched medium (Veltcheva *et al.*, 2005). High cytokinin to auxin ratios may modify the physiological state of the initial explants and initiate a somatic embryogenesis program.

The hormone free medium enhanced the development of somatic embryos through different stages (Fig. 5) and led to regeneration of complete plants (Pickardet *et al.*, 1989; Albrecht and Kohlenbach, 1989). Also the hormone free medium enhanced the formation of secondary embryos.

Here, within 8 months after the formation of the first embryo, 400 shoots per callus were developed from the somatic embryos derived from shoot tips of the cultivar Giza 2 on M1 treatment. However, these calluses maintained embryogenic potential over a period of 14 months. The number of shoots was increased due to the effect of the hormonal composition of the M1 treatment. Further, the type of explant used (shoot tip) with its endogenous hormones may have played an important role in this regeneration system. In the M2 treatment within 8 months only 127 shoots per callus were developed from the somatic embryos that derived from calluses of epicotyl of cultivar 24 Hyto. The difference between each cultivar in the number of shoots was varied due to the type of explants used and the hormonal composition of each treatment. Cultivar genetics seems to have an important effect on the regeneration rate.

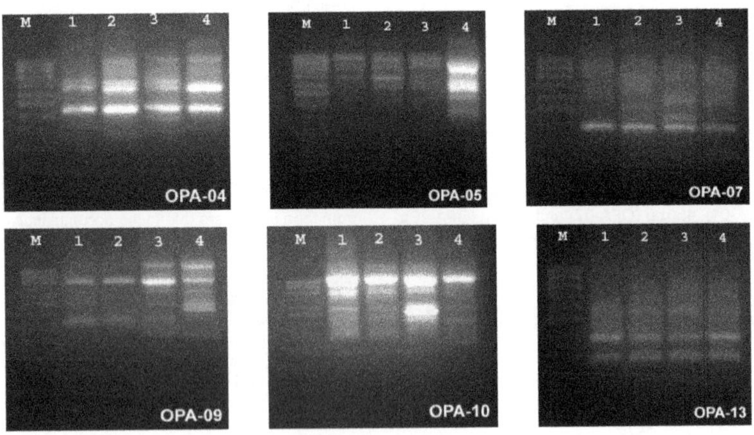

Fig. 4. RAPD profiles of faba bean cvs. were amplified with six RAPD primers. M: DNA molecular marker 100 bp. Lane 1: Control plant of cv. Giza 2 Lane 2: Regenerated plant of cv. Giza 2 via somatic embryogenesis . Lane 3: Control plant of cv. 24 Hyto. Lane 4: Regenerated plant of cv. 24 Hyto via somatic embryogenesis.

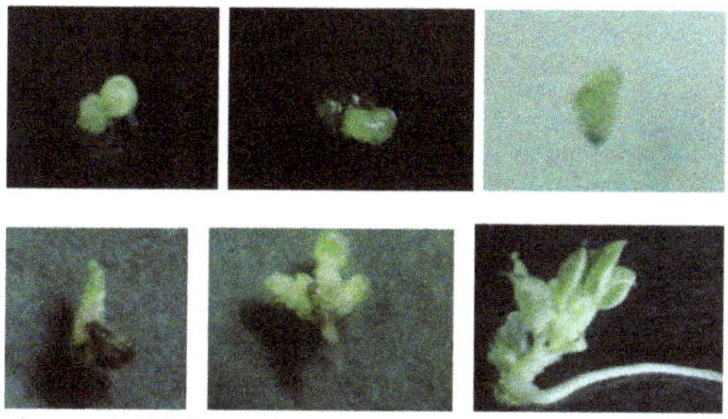

Fig. 5. Different stages of somatic embryogenesis (globular stage, heart stage, torpedo stage, cotyledonary stages, and young plantlets) of cv. Giza 2 were formed in the hormone free medium.

During the development of somatic embryos some of the developed plantlets formed roots rapidly in the same hormone free medium. However, sometimes the number of roots was up to 3 per plantlet (Fig. 6). A delay in the formation of roots was observed in some plantlets, in some cases the normal shoots were developed but root formation was suppressed.

Difficulties in the regeneration from the callus tissues of faba bean led to a limited number of publications on cell and tissue culture of this species. A regeneration system of faba bean through indirect somatic embryogenesis was reported by Griga *et al.* (1987). An indirect somatic embryogenesis system for *Vicia narbonensis* was reported by Pickardet *et al.* (1989) and also reported by Albrecht and Kohlenbach (1989). Another kind of regeneration is the organogenesis that was described in faba bean by Taha and Francis, (1990), Tegeder *et al.* (1995) and Böttinger *et al.* (2001).

Fig. 6. Plants developed from somatic embryogenesis of cv. Giza 2 with number of roots ranged from 1–3 roots.

PCR-based RAPD (Random amplified polymorphic DNA) markers have been applied to many types of genetic studies because of their simplicity and ease of use (Gutierrez *et al.*, 2006). In this study, RAPD analysis was performed to identify potential somaclonal variations among the regenerated plants from somatic embryogenesis. To investigate the effects of the media and subcultures, on genetic stability, the regenerated plants were obtained after long-term of subcultures (after about 13 months of the first culture, from the later generations).

More new bands were detected in the regenerated plants of cultivar Giza 2 while more bands were missing in regenerated plants of cultivar 24Hyto. Both the deletions underlying the loss of DNA fragments and the modifications underlying the appearance of new fragments among the regenerated plants may be used to help to improve faba bean genetically. Furthermore, RAPD analysis could be used to trace the genetic changes in the regenerated plants at the genome level. RAPD polymorphism in callus-derived plants has been reported in wheat by Brown *et al.*, 1993, and turmeric (Salvi *et al.*, 2001). However, Isabel *et al.* (1993) detected no polymorphism in regenerated plants of *Picea Mariana*. The source of the explants, mode of regeneration (somatic embryogenesis/organogenesis/axillary bud multiplication), genotypes used and the plant growth regulators applied are known to play a major role in determining the presence or absence of variations (Razdan, 1994; Damasco *et al.*, 1996). Several investigators (Link *et al.*, 1995; Torres *et al.*, 1993; Vaz Patto *et al.*, 1999; Gutierrez *et al.*, 2006; and Gutierrez *et al.*, 2007) used RAPD molecular markers to study the genetic variability and relationships among accessions, lines and cultivars of faba bean. Therefore, a link between natural and induced genetic variation might be made by further analyses.

Acknowledgments

We thank Dr. Said Abd-ELTawab (Cotton Research institute, Agricultural Research Center, Egypt) for his assistance with statistical analysis.

References

Albrecht, C., and Kohlenbach, H.W. (1989) Induction of somatic embryogenesis in leaf derived callus of *Vicia narbonensis* L. Plant Cell Rep. *8*, 267–269.

Anthony, J.M., Senaratna, T., Dixon, K.W., and Sivasithamparam, K. (2004) The role of antioxidants for initiation of somatic embryos with *Conostephium pendulum* (Ericaceae) Plant Cell Tiss. Org. Cult. *78*, 247–252.

Bond, D.A. (1987) Recent developments in breeding of field beans (*Vicia faba* L.) Plant Breed. *99*, 1–26.

Böttinger, P., Steinmetz, A., Schieder, O., and Pickardt, T. (2001) *Agrobacterium*-mediated transformation of *Vicia faba*. Mol. Breed. *8*, 243–254.

Brown P.T.H, Lange, F.D., Kranz, E., and Lörz, H. (1993) Analysis of single protoplasts and regenerated plants by PCR and RAPD technology. Mol. Gen. Genet. *237*, 311–317.

Damasco, O.P., Graham, G.C., Henry, R.J., Adkins, S.W., Smith, M.K., and Godwin, I.D. (1996) Random amplified polymorphic DNA (RAPD) detection of dwarf off-types in micropropagated Cavendish (*Musa* SPP AAA) bananas. Plant Cell Rep. *16*, 118–123.

Doyle, J.J., and Doyle, J.L. (1987) A rapid DNA isolation procedure for small quantities of fresh leaf tissue. Phytochem. Bull.*19*, 11–15.

Duke, J.A. (1981) Handbook of Legumes of World Economic Importance. (New York: Plenum Press), pp. 199–265.

Gamborg, O.L., Miller, R.A., and Ojima, K. (1968) Nutrient requirements of suspension cultures of soybean root cells. Exp. Cell Res. *50*, 151–158.

Griga, M., Kubalakova, M., and Tejklova, E. (1987) Somatic embryogenesis in *Vicia faba* L. Plant Cell Tiss. Org. Cult. *9*, 167–171.

Gutierrez, N., Avila, C.M., Duc, G., Marget, P., Suso, M.J., Moreno, M.T., and Torres, A.M. (2006) CAPs markers to assist selection for low vicine and convicine contents in faba bean (*Vicia faba* L.) Theor. Appl. Genet. *114*, 59–66.

Gutierrez, N., Avalia, C.M., Rodriguez-Suarez, C., Moreno, M.T., and Torres, A.M. (2007) Development of SCAR markers linked to a gene controlling absence of tannins in faba bean. Mol. Breed. *19*, 305–314.

Isabel, N., Tremblay, L., Michaud, M., Tremblay, F.M., and Bousquet, J. (1993) RAPDs as an aid to evaluate the genetic integrity of somatic embryogenesis-derived populations of *Picea mariana* (Mill.) B.S.P. Theor. Appl. Genet. *86*, 81–87.

Jelenić, S., Mitrikeski, P.T., Papeš, D., and Jelaska, S. (2000) *Agrobacterium*-mediated transformation of broad bean *Vicia faba* L. Food Technol. Biotechnol. *38*, 167–172.

Khalafalla, M., and Hattori, K. (2000) Ethylene inhibitors enhance *in vitro* root formation on faba bean shoots regenerated on medium containing thidiazuron. Plant Growth Regul. *32*, 59–63.

Kuchuk, N. (2001) Biotechnology. In: Carbohydrates in Grain Legume Seeds, C.L. Hedly, ed. (Wallingford, UK: CABI Publishing), pp. 145–207.

Link, W., Dixkens, C., Singh, M., Schwall, M., and Melchinger, A.E. (1995) Genetic diversity in European and Mediterranean faba bean germ plasm revealed by RAPD markers. Theor. Appl. Genet. *90*, 27–32.

Mohamed, M.F., Coyne, D.P., and Read, P.E. (1993) Shoot organogenesis in callus induced from pedicel explants of common bean (*Phaseolus vulgaris* L.) J. Amer. Soc. Hort. Sci. *118*, 158–162.

Murashige, T., and Skoog, F. (1962) A revised medium for rapid growth and bioassays with tobacco tissue cultures. Physiol. Plant. *15*, 473–497.

Pickardt, T., Huancaruna Perales, E., and Schieder, O. (1989) Plant regeneration via somatic embryogenesis in *Vicia narbonensis*. Protoplasma *149*, 5–10.

Razdan, M.K. (1994) An Introduction to Plant Tissue Culture (New Delhi: Oxford & IBH Publishing CO. PVT. LTD), pp. 225–243.

Salvi, N.D., George, L., and Eapen, S. (2001) Plant regeneration from leaf base callus of turmeric and random amplified polymorphic DNA analysis of regenerated plants. Plant Cell. Tiss. Org. Cult. *66*, 113–119.

Selva, E., Stouffes, B., and Briquet, M. (1989) In vitro propagation of *Vicia faba* L. by micro-cutting and multiple shoot induction. Plant Cell Tiss. Org. Cult. *18*, 167–179.

Taha, R.M., and Francis, D. (1990) The relationship between polyploidy and organogenetic potential in embryo and root-derived tissue cultures of *Vicia faba* L. Plant Cell Tiss. Org. Cult. *22*, 229–236.

Tegeder, M., Gebhardt, D., Schieder, O., and Pickardt, T. (1995) Thidiazuron-induced plant regeneration from protoplasts of *Vicia faba* cv. Mythos. Plant Cell Rep. *15*, 164–169.

Torres, A.M., Weeden, N.F., and Martín, A. (1993) Linkage among isozym, RFLP and RAPD markers in *Vicia faba*. Theor. Appl. Genet. *85*, 937–945.

Vaz Patto, M.C., Torres, A.M., Koblizkova, A., Macas, J., and Cubero, J.I. (1999) Development of a genetic composite map of *Vicia faba* using F_2 populations derived from trisomic plants. Theor. Appl. Genet. *98*, 736–743.

Veltcheva, M., Svetleva, D., Petkova, SP., and Perl, A. (2005) *In vitro* regeneration and genetic transformation of common bean (*Phaseolus vulgaris* L.)— Problems and progress. Scientia Horticulturae. *107*, 2–10.

Chapter 7

Cloning of a Novel Antifungal Promoter from *Phaseolus vulgaris* and the Determination of its Activity in Stably Transformed *Nicotiana tabacum* Plants

Eman A. Mahmoud, Solliman M. Mohei El-Din, Mourad A.M. Aboul-Soud, Ahmed M. Aboul-Enein, Ghanem A. Sobhy, and Hany A. El-Shemy

Abstract

To investigate the transcriptional regulation of gene expression, chimeric fusions, between the β-glucuronidase reporter gene (GUS) and the isolated promoter regions of the *pvPDF* gene (*pvPDF-PRO: GUS)*, were constructed and introduced into *Nicotiana tabacum.* Analysis of transgenic *pvPDF-PRO:GUS* tobacco plants indicated that GUS activity was observed with all the promoter constructs with the strongest being in leaf followed by stem and roots. These results clearly demonstrate that *pvPDF-PRO* is a strong inducible and near-constitutive promoter and emphasize the great application potential for plant genetic engineering studies. Interestingly, a search for putative *cis*-acting elements in the *pvPDF* promoter architecture revealed the presence of some important transcription regulatory elements including: CAAT Box, TATA Box, CATA Box, and light regulatory elements (CCA1, GATA, GT-1). Taken together, these results further our understanding of the regulation of the *pvPDF* promoter activity.

Introduction

Genetic transformation is a powerful tool for production of crop plants with increased resistance to phytopathogens. A number of transgenic cultivars with heightened tolerance to economically important pests and disease agents are in commercial production. However, in most cases the transgene is driven by a powerful constitutive promoter, such as the cauliflower mosaic virus *35S* (*CaMV35S*) and its derivatives, and is expressed at high levels even in the absence of pathogen invasion. Continuous synthesis and high accumulation of transgene products, especially toxins, could interfere with plant metabolic pathways and the overall expression of other valuable traits. In contrast, the use of promoters of plant defense genes has distinct advantages as the majority is only activated following plant attack by pests or pathogens. The use of native plant promoters can also help to avoid transgene silencing often associated with the presence of promoters of non-plant origin in the plant genome (Yevtushenko *et al.*, 2004; Vieweg *et al.*, 2004; Nishiuchi *et al.*, 2004; Ross *et al.*, 2004; Matarasso *et al.*, 2005; Rubio-Somoza *et al.*, 2006).

Plants have developed a variety of physical and biochemical defense barriers against pests and pathogens. Mechanical wounding of plant tissue, mimicking pathogen invasion or insect chewing, leads to the accumulation of mRNAs that encode proteins thought to be involved in plant defense (Yevtushenko *et al.*, 2004; Bowles *et al.*, 1990), and provides a convenient system to isolate and study defense-related genes and their upstream regulatory regions in transgenic hosts (Clarke *et al.*, 1994; Hollick *et al.*, 1993). Potato plants can be engineered for broad-spectrum disease resistance by expression of antimicrobial peptides under control of a constitutive *CaMV 35S* promoter (Osusky *et al.*, 2000).

In recent years, concerns over genetic modification issues have resulted in regulatory authorities requiring comprehensive analysis of transgene insertion events in the plants that are to be commercialized. Recent studies have suggested that applied and modified a genomic walking method that combines vectorette and suppression PCR walking. Some studies have suggested that stably expressed transgenes comprise relatively simple T-DNA arrangements flanked on at least one side by plant DNA and that unstably expressed loci tend to be composed of multiple T-DNA copies (Iglesias *et al.*, 1997). Several PCR-based walking methods have been described (Ochman *et al.*, 1988; Rosenthal and Jones, 1990; Riley *et al.*, 1990; Lagerstrom *et al.*, 1991; Parker *et al.*, 1991; Trueba and Johnson, 1996; Jones and Winistorfer, 1993). However, these methods generally have not been applied in determining T-DNA insertion sites because they are too inefficient and/or complicated.

The genes which encode the various components of the pre-mRNA splicing or rRNA processing machinery provide a rich source of promoters for transgene expression in plant biotechnology. As most genes are organised in multigene families with great variability in expression levels and patterns, a novel approach has been developed to allow the identification and isolation of promoters with the required expression characteristics. This approach proves valuable for promoter isolation and exploitation (Sunter and Bisaro, 2003; Van *et al.*, 2002).

This work describes the isolation and functional characterization of the plant-defensin 5′-untranscribed promoter from *Phaseolus vulgaris*, designated *pvPDF-PRO1*, *pvPDF-PRO2* and *pvPDF-PRO3* fragments. Their functional analyses and promoter activity in chimeric reporter constructs (*pvPDF-PRO: GUS*) in transgenic tobacco plants are discussed, in the light of the potential use of these promoters in plant biotechnology.

Materials and methods

Plant material and growth conditions
Kidney Bean, *Phaseolus vulgaris*, seeds were grown on MS media at 25±2°C for 7 days in dark. The harvested material was immediately frozen in liquid N_2 and stored at −70°C. For transformation studies, *Nicotiana tabacum* var Xanthi plants were grown at 25±2°C in 16h light and 8h dark.

Cloning strategy of kidney bean pvPDF *promoters*
The cloning of the *pvPDF* promoter was carried out according to Reddy *et al.* (1999 and 2002). Purified *Phaseolus vulgaris* genomic DNA was restriction-digested overnight at 37°C, with 80 units of restriction enzymes for the concentration of 40µg/l DNA: e.g. *Bam*HI, *Bgl*II and *Sou*3A (NEB). Following heat inactivation of enzymes at 65°C for 10 min and centrifugation, the resulting DNA pellets were then washed with 70% EtOH, allowed to air dry and subsequently resuspended in 20µl sterile double-distilled H_2O.The digested DNA were electrophoresed on 0.8% agarose gel and processed for cutting the region from 0.5 to 1.0kb. Each fragment was purified using Miniprep columns (Qiagen). The digested DNA (20µg) was partially end-filled by dGTP and dATP .The reaction mixture contained dNTPs (2mM each), 10X reaction buffer (2.5µl), klenow enzyme (2µl) in reaction volume of 50µl and was incubated at 37°C for 30min. The DNA was purified through Miniprep purification Columns (Qiagen). The following oligonucleotide adaptors were employed: ADOP-32, 5′-AATACGACTCACTATAGGGCGGCCGCCCGGGC-3′ and ADOP-27, 5′-CACTATACCCGCCGGCGGGCCCGCT-3 (Synthesized by IDT Inc.). A volume of 20µl of each adaptor (50ng/l) was pipetted into a 0.5-ml Eppendorf microfuge tube and overlaid with mineral oil. The adaptors were heated at 99°C for 4 min in a beaker of water. The heat was removed, and the solution was allowed to cool for 1 h at room temperature. The annealed adaptors were decanted from under oil and stored at −20°C. Then, 10µl of the genomic restriction digest was ligated to 1µl of the annealed adaptors with 2µl of T4 DNA ligase buffer and 2µl of T4 DNA ligase (5U/µl) in a 20-µl reaction. The ligation was incubated overnight at 12°C and heat inactivated at 65°C for 10 min. A 180-µl volume of TE (pH 8) was added to the ligation mix; this is called the adaptor library. Excess primer-adapter was removed by purification through QIAquik Columns (Qiagen).

Manipulation of plant nucleic acids
Nucleic acid manipulations were essentially carried out according to Sambrook *et al.* (1989). Genomic DNA was extracted and purified on a mini-prep scale as described by Murray and Thompson (1980). Total RNA from plant tissues was isolated as mentioned by Chomczynski and Sacchi (1987) with some modifications. Target PCR products were cleaned up by Gel Extraction Kit (Promega) and cloned into the pGEMT vector kit (Promega), then transformed into *E. coli* DH5α according to Hanahan (1985). Positive clones were confirmed by restriction analysis and sequenced using Sanger Sequencing Technology on ABI Prism 3730XL (Applied Biosystems/Sanger) according to the dideoxy chain-termination method (Sanger *et al.*, 1977).

Polymerase chain reaction
Taq-polymerase, dNTPs (deoxynucleotide triphosphate) and convergent primers achieved amplification of desired DNA fragments. The following adaptors were used to prime PCR reactions: ADOP-32, 5′-AATACGACTCACTATA GGGCGGCCGCCCGGGC-3′ and ADOP-27, 5′-CACTATACCCGCCGGCGGGCCCGCT-3. *pvPDF* promoter-based oligonucleotide primers were used: Def R1, 5′-CTCTTTATTCATCTCACTCGACT-3′; Def R2, 5′-ACAATTTTTGGTGCACCAACAACGAAATCAT-3′ and Def F1, 5′-AGTCGAGTGAGATGAATAAA GAGTTTTGAA-3. The reaction conditions for PCR involved denaturation at 94°C for 30 seconds, annealing at 52°C for 30 seconds and extension at 72°C for either 30 seconds (for *pvPDF-PRO* amplification) or 2 minutes (for *pvPDF-PRO::GUS*). After 30 cycles of amplification an aliquot of this reaction mixture was loaded onto a 0.8% agarose gel and checked. For nested PCR, the following primers were employed: Pro2-F-RE, 5′-GAGTTTTGAAAGGAGAATACTTTG AAGTTT-3′; Emandef-R1-RE, 5′-GGTGCACCAACAA CGAAATCAT-3′ and ProD-R-RE, 5′-GCCAGCTAG TGATTTCTTCTCCAT-3′.

In silico *sequence analyses*
The University of Wisconsin Genetic Computer group (GCG) sequence analysis software package, PC/GENE (Intelligenetics), and BLAST were utilized for sequence analyses (Hobohm and Sander, 1995; Stultz *et al.*, 1993; Altschul, 1997). For sequence manipulation BioEdit Sequence Alignment Editor was employed (Hall, 1999).

Preparation of chimeric constructs
Plasmid vector pBI121 (Bevan, 1984) was used for cloning of *pvPDF* promoter upstream from GUS gene (Fig. 1). These constructs were then utilized for tobacco transformation. To generate *pvPDF::GUS* promoter construct, *pvPDF* promoter fragments, double-digested with *Bam*H1/*Hind*III, were ligated into pBI121 binary vector digested with the same restriction enzymes. Ligation was carried out overnight at 14°C in the presence of T_4 DNA

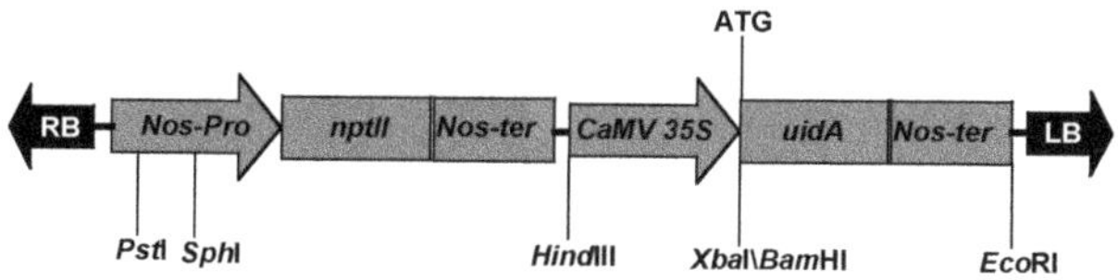

Fig. 1. The pBI121 binary vector used in chimeric construct generation including the map positions of some restriction sites.

ligase. The ligation reaction was employed to transforme DH5α competent cells.

Plant transformation

Constructs were transferred to *Agrobacterium tumefaciens* strain LBA4404 by the freeze-thaw method according to An *et al.* (1988). To test for *bona fide* transformants, 3-day-old *Agrobacterium* colonies were further analyzed by PCR for the presence of the selectable marker gene *nptII*, using the following primers combination: Neo5, 5′-GAGGCTATTCGGCTATGACTG-3′ and NeoSHR, 5′-GGCCATTTTCCACCATGATA-3′. Leaf discs of sterile-grown *Nicotiana tabacum* were transformed essentially as described by Horsch *et al.* (1985), with minor modifications. Transformed plants were selected by rooting several times on kanamycin-containing medium and transferred to compost.

GUS histochemistry and quantitative assay

This method for screening the transgenic plants was done according to (Jefferson, 1987 and Allan *et al.*, 1993) and allows for the verification of the expression of uid A gene in transgenic plants. Leaf tissue from wild type and transgenic plants was collected and rinsed in 50 mM Na-phosphate buffer (pH 7.0). Then the tissue was stained with 2 mM 5-bromo-4-chloro-3-indolyl glucuronide (X-gal from Biosynth. Inc) in 50 mM of Na-phosphate buffer (pH 7.0), followed by brief vacuum infiltration and placed at 37°C for overnight in dark. After staining, tissues were rinsed extensively in ethanol to remove chlorophyll before examination. Preparing X-GLUC Stain by Dissolve 5 mg X-gluc in 1.0 ml dimethyl formamide and adding 50 mM $NaPO_4$ pH 7.0 to 10 ml.

4-Methylumbelliferone (4-MU) production was measured at four time points using a fluorometer fitted with a microtiter plate reader (Millipore, Cytofluor 2350). Total protein determination was carried out according to Bradford (1976) using Bradford reagent ((Bio-Rad) and BSA as standard. Measurements were performed in triplicate and GUS activity expressed as pmol 4-MU/min/mg protein.

Results and discussion

Identification and cloning of pvPDF *promoters*

Plant promoter architecture is important for understanding regulation and evolution of the promoters, but our current knowledge about plant promoter structure, especially with respect to the core promoter, is insufficient. Several promoter elements including TATA box, and several types of transcriptional regulatory elements have been found to show local distribution within promoters, and this feature has been successfully utilized for extraction of promoter constituents from plant genome.

Promoters of various strengths and specificities are required for expression of foreign genes in plants for analysis of gene function or for biotechnological improvement of crop species. To meet such requirements and to have maximum precision in expression, a range of promoters with different expression levels and patterns is desired. Keeping this in mind, we carried out a program of isolation of plant promoters and functionally characterized them for suitability to express foreign genes using tobacco as a model system. One of the promoters, designated *pvPDF*, was identified and was shown to belong to a plant defensin gene encoding an antifungal protein from beans.

The isolation of unknown DNA sequences flanking known regions is critical for gene expression analysis. Several protocols have been developed to isolate an unknown DNA sequence (promoter) adjacent to DNA fragments of known sequence (cDNA) by PCR (Hui *et al.*, 1998). A number of modifications were developed to isolate the unknown 5′ and 3′ flanking regions of the DNA. Usually, PCR was carried out with restriction enzyme(s)-digested genomic DNA fragments after ligation (Siebert *et al.*, 1995) or cloning into a vector (Niu and Fallon, 1999) or ligated to double-stranded, partially double-stranded (Iwahana *et al.*, 1994; Willems, 1998), or single-stranded oligonucleotide cassettes (Kilstrup and Kristiansen, 2000). In these cases the amplifications were carried out with locus-specific primer(s) and a vector/oligonucleotide cassette specific-primer to amplify a fragment contiguous to the known sequence. In this method all the template molecules are likely to be amplified linearly leading to the generation of a lot of noise. In this study we have followed a protocol developed by Reddy *et al.* (2002) that involved the use of restriction digestion of genomic DNA followed by partial filling whereby preventing self-ligation between fragments. In addition, the present protocol also employed biotinilated primers that enrich specific template prior to nested PCR. Also, one of the adopter strands was blocked at the 3′ end by attaching an amine group (Reddy *et al.*, 2002). Following this method a number of unknown 3′ and 5′ regions of *pvPDF* gene promoters were isolated in this paper (data not shown). Namely, a 730 bp fragment corresponding to the 5′ UTR region of *pvPDF* gene from bean, was isolated following this new method (Fig. 2). The major transcript was found to initiate from 270 bases upstream of the translation

A

```
  -730      -720      -710      -700      -690      -680      -670
CTTTAGAATAAATAAAAATTTCAAATCCAAATATCTGTTTTATTATACTCTAAATACTAAAAAATATATG

  -660      -650      -640      -630      -620      -610      -600
TTAATTATAAACATTTATTTATTATTTTCTACAAAAAAGACTCTTTAAAAAAGATTAAAAAAATTCCTCT

  -590      -580      -570      -560      -550      -540      -530
ATTTTTTCATGTGCTCCCCTCCTCCAAACCTCTCCCCTATCCCTTCCCTACTTTTGTTTAATTACTCTTT

  -520      -510      -500      -490      -480      -470      -460
TATATATGAGATGCATATTATTGAATTTATTTTATCGATTACTTTTGGGAAAATTACATGATTAATTATC

  -450      -440      -430      -420      -410      -400      -390
ACAAACAAATCGAATGGAGAATTCAAATTGAATTGTTGCATACTATATTTGTCTGTCTCCATATATAGCA

  -380      -370      -360      -350      -340      -330      -320
TTAATTATATGTTGATAAGCTCCACTCTAGCTTCCACGCCCCACACTATTGACTTTTCATCATGAAAATA

  -310      -300      -290      -280      -270      -260      -250
CAAACTTGGGTAATAAAATATGCTGCCACATCAATCCGTTTCTTGTGGTTTCTTCTAGCTTCTCAAACTT

  -240      -230      -220      -210      -200      -190      -180
CATTTTCAGATTAATAATAATATTTGTTCATTTCTCTTACTAGTCAAGCCCCGGCCAAAAAAGGGAAGGG

  -170      -160      -150      -140      -130      -120      -110
GAAGAAATCGCTTTTGGCCCTTTTTTTGGGGGGGGGGACCCAAAAAAAAAAAATGGCAGCCAAGAGTTT

  -100      -90       -80       -70       -60       -50       -40
TTTTTCCCTTCCCCTTTTTGTGGCGCCAAAAACACAACAAAATTAATGATATGTTTATATAGTCTCTTTA

  -30       -20       -10        0        10        20
ATAGATCATTGTTACTGCTGAATACTTTATAATTATGGACAAGAAATCACTAGCTGGC
```

B

DEF F1 primer

```
ATGGACAAGAAATCACTACCTGGCTTATGCTTCCTCTTCCTTGTTCTCTT
TGTTGCTCGTAAGTCACTACTCTTATGTTTATTTTTCTGCATGTCTATAA
CTTTAATCATTCTCCTTACATCCTATAACTGTCTATTTACATATTCAGTG
TAACTTTTAGTTTTCTACAAGATTAATGTACTGTTTATACAGTCTCTTTA
GTTGTTCATTGTTACTGCTGAATACTTTATAATTCAATATCCCCGATATA
TGTTGATGAATTTATTTGAACATGCAGAAGAATGTGTGTTGCAGACAGAG
GCAAAGACTTGCGAGAACCTGGCTGATACATACAGGGGTCCATGCTTCAC
CACTGGAAGCTGCGATGATCACTGCAAGAACAAAGAACACTTGCTGAGTG
GCAGATGCAGGGATGATTTCGTTGTTGGTGCACCAAAAATTGTTAA
```

DEF R1 primer

Fig. 2. **A** The nucleotide sequence of the cloned *pvPDF* promoter generated by BioEdit Sequence Alignment Editor (Hall, 1999). **B** The nucleotide sequence of the *pvPDF* full length gene after isolated from Kidney Beans and sequencing.

initiation site. The minor transcript was found to initiate from 150 nucleotides upstream of the translation initiation site (data not shown). Similar result has been obtained by Reddy *et al.* (1999), in which they isolated genomic DNA fragments from tobacco- and pea-derived promoter regions for DNA topoisomerase I (*topo I*) using similar PCR based 5′ genome walking. In case of pea, they isolated 1140 bp and in tobacco a 482 bp upstream of ATG -5′ flanking region of tobacco *topo I* using 5′-RACE PCR-based approach (Reddy *et al.*, 1999).

Moreover, we have managed to cloned a putative 5′-UTR promoter region for the isolated *pvPDF* gene and have found that it possess various known plant regulatory motifs (Table 1, Fig. 2 and Fig. 3A). Futhermore, the corresponding *pvPDF* gene was also successfully PCR-amplified (Fig. 3 B). The *pvPDF* gene open reading frame was found to be composed of 195 bp and the exact sequence will be submitted to GenBank (Mahmoud *et al.*, unpublished data).

Database-assisted pvPDF *promoter sequence analyses*

The sequence homology of the 5′-UTR region of the *pvPDF* gene isolated in the present study (Fig. 2) revealed the existence of some differences within the reported *pvPDF* gene sequence (data not shown). This could be due to the fact that the *pvPDF* exists in multiple copies and the difference could be due to isolation of promoter for the gene other than the one that was previously reported by (Siva Reddy ICGEB-India, personal communication). Analysis of regulatory sequences is greatly facilitated by database-assisted bioinformatic approaches. In this context, the TRANSFAC database contains information on transcription factors, their origin, functional properties and their sequence-specific binding activities (http://www.sphinx.rug.ac.be: 8080/PLANTCARE/) (Hobohm and Sander, 1995; Stultz *et al.*, 1993).

By employing software tools it is possible to screen the database with a given DNA sequence for interacting transcription factors. If a regulatory function is already attributed to this sequence the database assisted identification of binding sites for proteins or protein classes and subsequent experimental verification may establish functionally relevant sites within this sequence. The binding transcription factors as well as interacting factors may already be present in the database. The putative cis-acting elements present in the isolated *pvPDF* promoter were identified using PLACE (Higo *et al.*, 1999, http://www.dna.affrc.go.jp) (Table 1). A search for putative *cis*-acting elements in the *pvPDF* promoter revealed the presence of some important elements including: CAAT Box, TATA Box, CATA Box, and light regulatory elements or LREs (CCA1, GATA, GT-1). Interestingly, the *pvPDF* promoter has a TATA Box near the 5′ end of the *pvPDF* gene. The first TATA-like element detected is at −237 and the second TATA Box was at −685. The presence of CAAT Box1 motifs in the *pvPDF* promoter correlates with higher expression in leaves as compared to roots. Similar motifs have been reported in the promoter of the *legA* gene of pea (Shirsat et.al. 1989). The presence of light regulatory elements has the consensus GT1 motif binding site in many light-regulated genes (Le Gourrierec et.al. 1999). In the *pvPDF* promoter one light regulatory element could been detected at the region between −251 to −258. Moreover, two GATA Box elements were identified, one at −678 to −682 and the other between −320 to −325. In this context, similar *cis*-acting elements have been reported in CaMV 35S promoter and in promoters from Petunia (Benfey and Chua, 1990; Gilmartin *et al.*, 1990). The presence of light regulatory elements and tissue specific elements accounts for the tissue specific light regulation of *pvPDF*. Three putative Dof binding sites are present in the region between −703 to −706. Dof proteins are unique to plants and contain a highly conserved DNA

Table 1. Putative *cis*-acting elements identified in the cloned *pvPDF* promoter sequence. These elements were identified using the Signal Scan Program at PLACE (http://www.dna.affrc.go.jp).

Reference	Sequence	Site Position	Found in	Motif
Shirsat *et al.* (1989)	CAAT	452	Sequences responsible for the tissue specific promoter activity of a pea legumin gene	CAAT BOX1
Rubio-Somoza *et al.* (2006)	GATA	678	GATA motif in CaMV 35S promoter	GATA BOX
Matarasso *et al.* (2005)	TAAAATAT	434	chlorophyll a/b binding protein the promoter of Petunia	LECPLEACS2
Vieweg *et al.* (2004)	CTCTT	524	One of the consensus sequence motifs of organ-specific elements	OSE2 ROOT NOD-ULE
Tjaden *et al.* (1995)	TTATTT	685	5′ UTR region of pea glutamine synthetase gene	TATA BOX5
Nishiuchi *et al.* (2004)	TGACY	400	"W box" found in the promoter region of a transcriptional repressor ERF3 gene in tobacco; May be involved in activation of ERF3 gene by wounding	W BOX NTERF3
Tjaden *et al.* (1995)	TATAAAT	237	5′ UTR region of pea *legA* gene	TATA BOX2
Ross *et al.* (2004)	NGATT	498	ARR1; Response regulator	ARR1 AT
Elmayan and Tepfer (1995)	ATATT	225	Motif found both in promoters of rolD	ROOT MOTIF TAPOX1
Zhou (1999)	GRWAAW	251	Consensus GT-1 binding site in many light-regulated genes,	GT-1 CONSENSUS
Le Gourrierec *et al.* (1999)	GRWAAW	429	GT-1 binding site in many light-regulated genes	GT-1 CONSENSUS

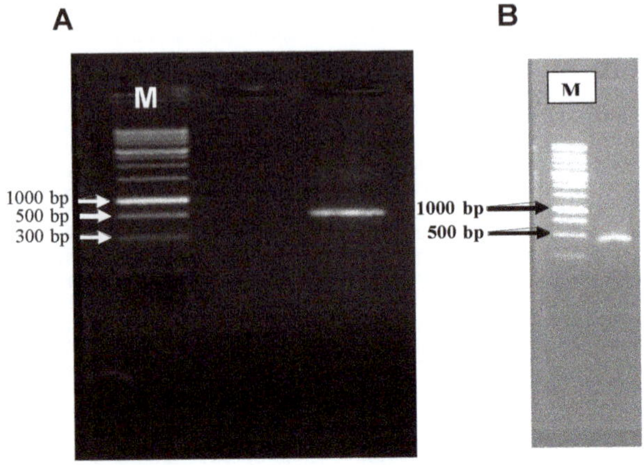

Fig. 3. **A** Photograph of a gel showing a purified PCR product corresponding to the 730 bp full-length promoter of *pvPDF* from Kidney Beans prior to pGEMT cloning and sequencing. M,marker lane. **B** PCR-based isolation of the coding region of the *pvPDF* gene from Kidney Beans using the full-length primer set (See Materials and Methods).

binding domain that binds to a core AAAG sequence. In maize, Dof1 is constitutively expressed in roots, leaves and stem and acts as a transcriptional activator while Dof2 is expressed in roots and stem and acts as a transcriptional repressor (Yanagisawa and Sheen, 1998) for light mediated expression of C4 photosynthetic phosphoenolpyruvate carboxylase (C4-*PEPC*). Thus, Dof proteins may have tissue specific effects. Dof proteins have also been implicated in regulation of hormonal responses and pathogen attack (Yanagisawa and Schmidt, 1999; Yanagisawa, 2000). The *pvPDF-PRO* promoter sequence also has a "Box II" box sequence at 231 to 234 sites. Similar *cis*-acting elements have been reported to be present in the tobacco plastid *atpB* gene promoter (Kapoor and Sugiura, 1999).

Responsiveness of *Promoter Transgenic* pvPDF: GUS *Tobacco Plants*

Agrobacterium mediated leaf disk method was followed for transformation, *pvPDF* promoter sequences and the GUS expression. A frequency of 5–20 plantlets were obtained per explants, which was comparable to control pBI121 binary vector transformation frequency, indicating that the cloning of *pvPDF* promoter had no adverse effects on the overall transformation frequencies. Transformed plants rooted normally and no abonormalities associated with the transformation was observed (Fig. 4).

Histochemical GUS staining of transgenic plants indicated that pvPDF: GUS promoter was active in almost all reproductive organs and the highest level was in the roots (Fig. 5).

We have utilized *Hind*III and *Bam*HI sites present in the pBI121 vector to clone *pvPDF* promoters. Three different constructs were created that differed in their length. The *pvPDF-PRO1* is the smallest promoter which was only 350 bp long, the *pvPDF-PRO2* was 500 bp and the longest promoter, *pvPDF-PRO3*, contained 730 bp. All the promoters were cloned into the same *Hind*III and *Bam*HI sites (Fig. 1) for a better comparison of promoter strength. As a control, pBI121 having 35S promoter was used. The CaMV 35S is a strong and constitutive promoter and used most extensively to express foreign genes in plants. Under 35S promoter, GUS expression was detected in all tissue types and at all developmental stages

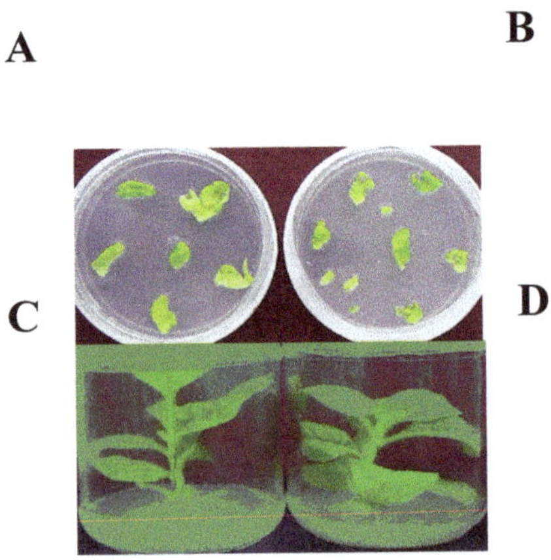

Fig. 4. Steps followed in the generation of transgenic *pvPDF::GUS* plants. **A** Tobacco explants after infecting the wounded plant tissue with *Agrobacterium* carrying the binary vector pBI121. **B** Transformed cells divide and grow. **C** Shooting and elongation medium. **D** Shoots are placed on another medium to promote root development, so that complete plantlets are produced.

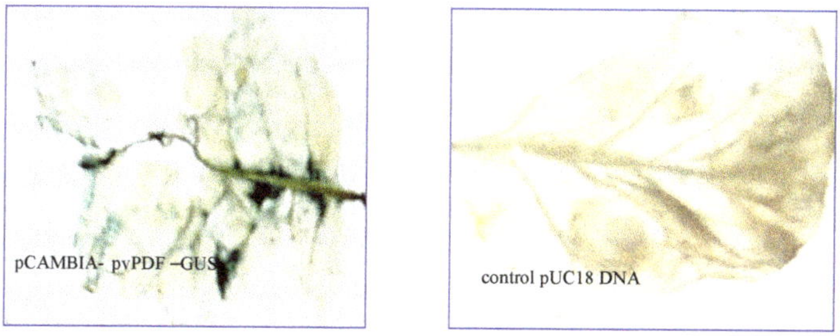

Fig. 5. Tobacco leaf was *bombarded* with plasmid DNA for transient assay of GUS. No blue spots could be seen on control leaf bombarded with pUC18 DNA. **A** pBI121 – pvPDF – GUS. **B** pUC18 DNA .

of tobacco plant growth. The expression was strongest in leaf tissue. The expression of GUS was observed with all three promoter constructs. The expression was strongest in the leaf followed by stem and roots (data not shown). Among the three constructs, was the highest *pvPDF-PRO3* followed by *pvPDF-PRO2*, whereas GUS activity was low with *pvPDF-PRO1* promoter construct (data not shown). However, the expression pattern for various tissues was similar in all the constructs.

The activity of the *pvPDF* gene promoter, encoding a bean cysteine-rich antifungal peptide, was investigated in transgenic *pvPDF::GUS* tobacco plants. Quantitative GUS activity analysis of the transgenic plant leaves showed the average activity of the bean *pvPDF-PRO* was 2- to 3-fold higher than that of the *CaMV35S* promoter. Histochemical GUS staining of transgenic plants indicated that *pvPDF::GUS* promoter was active in almost all reproductive organs and the highest level was in the roots (data not shown).

Functional analysis of promoter 5′-deletion series indicated that promoter activity of a 355 nucleotide fragment (–355 to the transcription initiation site) and a 460 nucleotide fragment (–460 to the transcription initiation site) were 2-fold and 3-fold stronger than that of the *pvPDF* full-length promoter, respectively. These results demonstrate that the bean *pvPDF* promoter is a strong inducible and near-constitutive promoter in plants and has great application potential for plant genetic engineering studies. Although the *CaMV35S* promoter appeared to be a strong, constitutive promoter in assays involving cell extracts, detailed histological analysis of a reporter gene product that is detectable at the cell and tissue level showed a rather high degree of variability of expression of this gene product. This histological analysis revealed an unknown and unexpected variability in the expression of a gene product driven by the CaMV35S promoter. This variable level and site of expression is believed to have two primary causes, one reason is attributed to the fact that variability is an intrinsic property of the CaMV35S promoter (Rubio-Somoza *et al.*, 2006; Matarasso *et al.*, 2005; Vieweg *et al.*, 2004; Nishiuchi *et al.*, 2004; Tjaden *et al.*, 1995; Ross *et al.*, 2004).

Promoters of various strengths and specificities are required for expression of foreign genes in plants for functional analysis of gene expression and for biotechnological improvement of economically-important crop plants. To meet such requirements, a range of promoters with different expression levels and patterns is required. Keeping this in mind, we begun a program of isolation of plant promoters and characterize them for their suitability to express foreign genes using tobacco as a model system. In this work, we described the cloning and functional analysis of a *Phaseolus vulgaris* promoter (*pvPDF*), controlling the expression of a plant defensin gene encoding an antifungal protein. Based on the obtained results, we envisage that this promoter can be valuable for various plant biotechnological applications.

References

Allan, G.C., Hall, G.E., Childs, L.C., Weissinger, A.K., Spiker, S., and Thompson, W.F. (1993) Scaffold attachment regions increase reporter gene expression in stably transformed plant cells. Plant Cell *5*, 603–613.

Altschul, S.F., Madden, T.L., Schaffer, A.A., Zhang, J., Zhang, Z., Miller, W., and Lipman, D.J. (1997) Gapped BLAST and PSI-BLAST: a new generation of protein database search programs. Nucleic Acids Res. *25*, 3389–3402.

An, G., Ebert, P.R., Mitra, A., and Sam, B.H. (1988) Binary vectors. Plant Mol. Biol. Manual (A3): 1–19.

Benfey, P.N. and Chua, N.H. (1990) The cauliflower mosaic virus 35S promoter: combinatorial regulation of transcription in plants. Science *250*, 959–966.

Bevan, M.W. (1984) Binary *Agrobacterium* vectors for plant transformation. Nucleic Acids Res. *12*, 8711–8721.

Bowles, D.J. (1990) Defense related proteins in higher plants. Annu. Rev. Biochem. *59*, 873–907.

Bradford, M.M. (1976) A rapid and sensitive method for the quantitation of microgram quantities of protein utilizing the principle of protein-dye binding. Anal. Biochem. *72*, 284–254.

Chomczynski, P. and Sacchi, N. (1987) Single-step method of RNA isolation by acid guanidinium thiocyanate-phenol-chloroform extraction. Anal. Biochem. *162*, 156–159.

Clarke, H.R., Davis, J.M., Wilbert, S.M., Bradshaw, H.D., and Gordon, M.P. (1994) Wound-induced and developmental activation of a poplar tree chitinase gene promoter in transgenic tobacco. Plant Mol. Biol. *25*, 799–815.

Dominguez, O., and Lopez-Larrea, C. (1994) Gene walking by unpredictably primed PCR. *Nucleic Acids Res. 22*, 3247–3248.

Elmayan, T., and Tepfer, M. (1995) Evaluation in tobacco of the organ specificity and strength of the rol D promoter, domain A of the 35S promoter and the 35S^2 promoter Transgenic Res. *4*, 388–396.

Gilmartin, P.M., Sarokin, L., Memelink, J., and Chua N.-H. (1990) Molecular light switches for plant genes. Plant Cell *2*, 369–378.

Hall, T.A. (1999) BioEdit: a user-friendly biological sequence alignment editor and analysis program for Windows 95/98/NT. Nucl. Acids Symp. Ser. *41*, 95–98.

Hanahan, D. (1985) Heritable formation of pancreatic β-cell tumours in transgenic mice expressing recombinant insulin/simian virus 40 oncogenes. Nature *315*, 115–122.

Higo, K., Ugawa, Y., Iwamoto, M., and Korenaga, T. (1999) Plant cis-acting regulatory DNA elements (PLACE) database: Nucleic Acids Res. *27*, 297–300.

Hobohm, U. and Sander, C. (1995) A sequence property approach for sequencing protein databases. J. Mol. Biol. *251*, 390.

Hollick, J.B. and Gordon, M.P. (1993) A poplar tree proteinase inhibitor-like gene promoter is responsive to wounding in transgenic tobacco, Plant Mol. Biol. *22*, 561–572.

Horsch, R.B., Fry, J.F., Hoffman, N.L., Eichholtz, D., Rogers, S.G., and Fraley, R.T. (1985) A simple and general method for transferring genes into plants. Science *227*, 226–228.

Hui, E.K., Wang, P.C., and Lo, S.J. (1998) Strategies for cloning unknown cellular flanking DNA sequences from foreign integrants. Cell Mol. Life Sci. *54*, 1403–1411.

Iglesias, V.A., Moscone, E.A., Papp, I., Neuhuber, F., Michalowski, S., Phelan, T., Spiker, S., Matzke, M., and Matzke, A.J. (1997) Molecular and cytogenetic analyses of stably and unstably expressed transgene loci in tobacco. Plant Cell *9*, 1251–1264.

Iwahana, H., Tsujisawa, T., Katashima, R., Yoshimoto, K., and Itakura, M. (1994) PCR with end trimming and cassette ligation: A rapid method to clone exon-intron boundaries and a 5′-up-stream sequences of genomic DNA based on a cDNA sequence. PCR Methods Appl. *4*, 19–45.

Jefferson, R.A., Kavanagh, T.A., and Bevan, M.W. (1987) GUS fusions: beta-glucuronidase as a sensitive and versatile gene fusion marker in higher plants. EMBO J. *6*, 3901–3907.

Jensen, J.S., Marcker, K.A., Otten, L., and Schell J. (1986) Nodule-specific expression of a chimeric soybean leghemoglobin gene in transgenic *Lotus comiculatus*. Nature *321*, 669–674.

Jones, D.H. and Winistorfer, S.C. (1993) Genome walking with 2- to 4-kb steps using panhandle PCR. PCR Methods Appl *2*, 197–203.

Kapoor, S. and Sugiura, M. (1999) Identification of two essential sequence elements in the nonconsensus type, I.I. PatpB-290 plastid promoter by using plastid transcription extracts from cultured tobacco BY-2 cells. Plant Cell *11*, 1799–1810.

Kilstrup, M. and Kristiansen, K.N. (2000) Rapid genome walking: A simplified oligo-cassette mediated polymerase chain reaction using a single genome-specific primer. Nucleic Acids Res. *28*, 55.

Lagerstrom, M., Parik, J., Malmgren, H., Stewart, J., Pettersson, U., and Landegren, U. (1991) Capture PCR: Efficient amplification of DNA fragments adjacent to a known sequence in human and YAC DNA. PCR Methods Appl. *1*, 111–119.

Le Gourrierec, J., Li, Y.F., and Zhou, D.X. (1999) Transcriptional activation by Arabidopsis GT-1 may be through interaction with TFIIA-TBP-TATA complex. Plant J. *18*, 663–668.

Matarasso, N., Schuster, S. and Avni, A. (2005) A novel plant cysteine protease has a dual function as a regulator of 1-aminocyclopropane-1-carboxylic Acid synthase gene expression. Plant Cell *17*, 1205–1216.

Murray, M.G. and Thompson, W.F. (1980) Rapid isolation of high molecular weight plant DNA. Nucleic Acids Res. *8*, 4321–4325.

Nishiuchi, T., Shinshi, H., and Suzuki, K. (2004) Rapid and transient activation of transcription of the ERF3 gene by wounding in tobacco leaves: Possible involvement of NtWRKYs and autorepression. J. Biol. Chem. *279*, 55355–55361.

Niu, L.L., and Fallon, A.M. (1999) Retrieval of flanking DNA using a PCR-based approach with restriction enzyme-digested genomic DNA template. Biotechniques *26*, 630–634.

Ochman, H., Gerber, A.S., and Hartl, D.L. (1988) Genetic applications of an inverse polymerase chain reaction. Genetics *120*, 621–623.

Odell, J.T., Nagy, F., and Chua, N.H. (1985) Identification of DNA sequences required for activity of the cauliflower mosaic virus 35S promoter. Nature *313*, 810–812.

Osusky, M., Zhou, G., Osuska, L., Hancock, R.E., Kay, W.W., and Misra, S. (2000) Transgenic plants expressing cationic peptide chimeras exhibit broad-spectrum resistance to phytopathogens, Nat. Biotechnol. *18*, 1162–1166.

Parker, J.D., Rabinovitch, P.S., and Burmer, G.C. (1991) Targeted gene walking polymerase chain reaction. Nucleic Acids Res. *19*, 3055–3060.

Reddy, M.K., Nair, S., and Sopory, S.K. (2002) A new approach for efficient directional genome walking using Polymerase Chain Reaction. Anal. Biochem. *306*, 154–158.

Reddy, M.K., Nair, S., Tewari, K.K., Mudgil, Y., Yadav, B.S., and Sopory, S.K. (1999) Cloning and characterization of a cDNA encoding topoisomerase, I.I. in pea and analysis of its expression in relation to cell proliferation. Plant Mol. Biol. *41*, 125–137.

Riley, J., Butler, R., Ogilvie, D., Finniear, R., Jenner, D., Powell, S., Anand, R., Smith, J.C., and Markham, A.F. (1990) A novel, rapid method for the isolation of terminal sequences from yeast artificial chromosome (YAC) clones. Nucleic Acids Res. *18*, 2887–2890.

Rosenthal, A. and Jones, D.S. (1990) Genomic walking and sequencing by oligo-cassette mediated polymerase chain reaction. Nucleic Acids Res. *18*, 3095–3096.

Ross, E.J., Stone, J.M., Elowsky, C.G., Arredondo-Peter, R., Klucas, R.V., Sarath G. (2004) Activation of the Oryza sativa non-symbiotic haemoglobin-2 promoter by the cytokinin-regulated transcription factor, ARR1. J. Exp. Bot. *55*, 1721–1731.

Rubio-Somoza, I., Martinez, M., Abraham, Z., Diaz, I., and Carbonero, P. (2006) Ternary complex formation between HvMYBS3 and other factors involved in transcriptional control in barley seeds. Plant J. *47*, 269–281.

Sambrook, J., Fritsch, E.F., and Maniatis, T. (1989) Molecular Cloning: A laboratory manual. 2 [1, *2*, 3]. Cold Spring Harbour, New York, Cold Spring Harbour Laboratory Press.

Sanders, P.R., Winter, J.A., Barnason, A.R., Rogers, S.G., and Fraley, R.T. (1987) Comparison of cauliflower mosaic virus 35S and nopaline synthase promoters in transgenic plants. Nucleic Acids Res. *15*, 1543–1558.

Sanger, F., Nicklem, S., and Coulson, A.R. (1977) DNA sequencing using chain termination inhibitors. Proc. Natl. Acad. Sci. U.S.A. *74*, 5463–5467.

Siebert, P.D., Chenchik, A., Kellogg, D.E., Lukyanov, K.A., and Lukyanov, S.A. (1995) An improved PCR method for walking in uncloned genomic DNA. Nucleic Acids Res. *23*, 1087–1088.

Silver J. and Keerikatte, V. (1989) Novel use of polymerase chain reaction to amplify cellular DNA adjacent to an integrated provirus. Virology *63*, 1924–1928.

Shirsat, A., Wilford, N., Croy, R., and Boulter, D. (1989) Sequences responsible for the tissue specific promoter activity of a pea legumin gene in tobacco. Mol. Gen. Genet. *215*, 326–331.

Stultz, C.M., White, J.V., and Smith, T.F. (1993) Structure analysis based on state-space modeling. Protein Sci. *2*, 305.

Sunter, G. and Bisaro, D.M. (2003) Identification of a minimal sequence required for activation of the tomato golden mosaic virus coat protein promoter in protoplasts. Virology *305*, 452–462.

Tjaden, G., Edwards, J.W., and Coruzzi, G.M. (1995) Cis elements and trans-acting factors affecting regulation of a nonphotosynthetic light-regulated gene for chloroplast glutamine synthetase. Plant Physiol. *108*, 1109–1117

Triglia, T., Peterson, M.G., and Kemp, D.J. (1988) A procedure for *in vitro* amplification of DNA segments that lie outside the boundaries of known sequences. Nucleic Acids Res. *16*, 8186.

Trueba, G.A. and Johnson, R.C. (1996) Random primed gene walking PCR: a simple procedure to retrieve nucleotide fragments adjacent to known DNA sequences. Biotechniques *21*, 20.

Willems, H. (1998) Adaptor PCR for the specific amplification of unknown DNA fragments. Biotechniques *24*, 26–28.

Van, W.R., Dong, X., Liu, H., Tien, P., Stanley J., and Hong, Y. (2002) Mutation of three cysteine residues in tomato yellow leaf curl virus-China C2 protein causes dysfunction in pathogenesis and posttranscriptional gene-silencing suppression. Mol. Plant Microbe Interact. *15*, 203–208.

Vieweg, M.F., Fruhling, M., Quandt, H.J., Heim, U., Baumlein, H., Puhler, A., Kuster, H., and Andreas, M.P. (2004) The promoter of the *Vicia faba* L. leghemoglobin gene VfLb29 is specifically activated in the infected cells of root nodules and in the arbuscule-containing cells of mycorrhizal roots from different legume and nonlegume plants. Mol. Plant Microbe Interact. *17*, 62–69.

Yanagisawa S and Sheen J. (1998) Involvement of maize Dof zinc finger protein in tissue-specific and light regulated gene expression. Plant Cell *10*, 75–89.

Yanagisawa S and Schmidt, R.J. (1999) Diversity and similarity among recognition sequences of Dof transcription factors. Plant J. *17*, 209–214.

Yanagisawa, S. (2000) Dof1 and Dof2 transcription factors are associated with expression of multiple genes involved in carbon metabolism in maize. Plant J. *21*, 281–288.

Yevtushenko, D.P., Sidorov, V.A., Romeroa, R., Kaya, W.W., and Misra, S. (2004) Wound-inducible promoter from poplar is responsive to fungal infection in transgenic potato. Plant Sci. *167*, 715–724.

Zhou, D.X. (1999) Regulatory mechanism of plant gene transcription by GT-elements and GT-factors.Trends Plant Sci. *4*, 210–214.

Chapter 8

The Interactions of the Largest Subunit of RNA Polymerase II with Other Cellular Proteins: a Bioinformatic Approach

Abhijit Shukla, Aparna Natarajan, Sukesh Bhaumik, Hany A. El-Shemy, and David A. Lightfoot

Abstract

The function of a protein is governed by its interaction with other proteins inside a cell. Therefore, it is important to identify the interacting partners of a particular protein to decipher its function. The protein interaction networks are generally determined by bioinformatic as well as experimental methodologies such as yeast two hybrid, mass spectrometry, immunoprecipitation, and fluorescence resonance energy transfer assays. Here, we have analyzed bioinformatically the interactions of Rpb1p (the largest subunit of RNA Polymerase II) with other proteins in yeast, using *Cytoscape* software and *Biogrid/ Biomart* database. We find that Rpb1p interacts with a large number of proteins involved in mRNA synthesis, processing, export, and other cellular processes. These results validate the application of such bioinformatic approach to determine the interactome for other cellular proteins.

Introduction

All cellular processes are carried out by the concerted actions of proteins through specific interaction networks. Therefore, protein function annotation has become an important area of research in post-genomic sequencing era. Thus, an understanding of various cellular processes demands elucidation of the protein-protein interaction networks within cell. Several experimental methodologies such as yeast two hybrid, mass spectrometry, immunoprecipitation, and fluorescence resonance energy transfer assays have been employed to decipher cellular protein-protein interactions. Based on these experimental studies, several bioinformatic tools have been developed to comprehensively analyze protein interaction networks of different cellular proteins. Here, we have used the Cytoscape software (Zhang *et al.*, 2007) and Biogrid/ Biomart database to identify the interactions of the largest subunit of RNA Polymerase II (RNAPII), Rpb1, with other proteins in yeast. Such analysis has revealed a large number of primary interactions of Rpb1p with many proteins involved in the regulation of transcription, chromatin structure, DNA repair, and other cellular events as discussed below.

Rpb1 and its interactions with other RNAPII subunits

The protein coding genes are transcribed into mRNA by RNAPII that is highly conserved from yeast to human. RNAPII is composed of 12 different subunits. These subunits are Rpb1, Rpb2, Rpb3, Rpb4, Rpb5, Rpb6, Rpb7, Rpb8, Rpb9, Rpb10, Rpb11, and Rpb12. Rpb1 is the largest subunit, and is essential to maintain the structural integrity of RNAPII. Moreover, it has Mg^{2+}-dependent polymerase activity. The Rpb1 and Rpb2 subunits are located at the core of RNAPII with smaller subunits at the surface (Cramer *et al.*, 2001, 2008; Bushnell and Kornberg, 2003; Cramer, 2004). Thus, Rpb1 interacts with several RNAPII subunits as also observed in this bioinformatic analysis (Fig. 1).

An important feature of Rpb1 is its c-terminal domain (CTD) that consists of multiple heptapeptide repeats (YSPTSPS) (Buratowski, 2003; Egloff and Murphy, 2008). Serine-2 and serine-5 (S-2 and S-5) of the heptapeptide in the CTD are phosphorylated and dephosphorylated during transcription by different kinases and phosphatases. Such phosphorylation status of Rpb1-CTD plays an important role in integrating various nuclear events through a variety of proteins involved in mRNA synthesis, processing, and export (Buratowski, 2003; Egloff and Murphy, 2008; Pandit *et al.*, 2008). In addition, Rpb1 interacts with Mediator, general transcription factors and activator, chromatin modifying and remodeling factors, and several other proteins to maintain normal cellular functions. These interactions are discussed below.

Interaction of Rpb1 with Mediator

The yeast Mediator complex comprises of a large number of subunits that include Srb (suppressor of RNA polymerase B) proteins (Srb2, Srb4, Srb5, Srb6, Srb7, Srb8, Srb9, Srb10, Srb11), Med proteins (Med1, Med2, Med4, Med6, Med7, Med8 and Med11), and several other polypeptides (Pgd1, Cse1, Nut2, Rgr1, Gal11, Sin4, Rox3 and Soh1) (Kornberg, 2005). The core of the Mediator complex (known as core mediator complex) lacking Srb8, Srb9, Srb10 and Srb11 (or the Srb8–11 module) has a

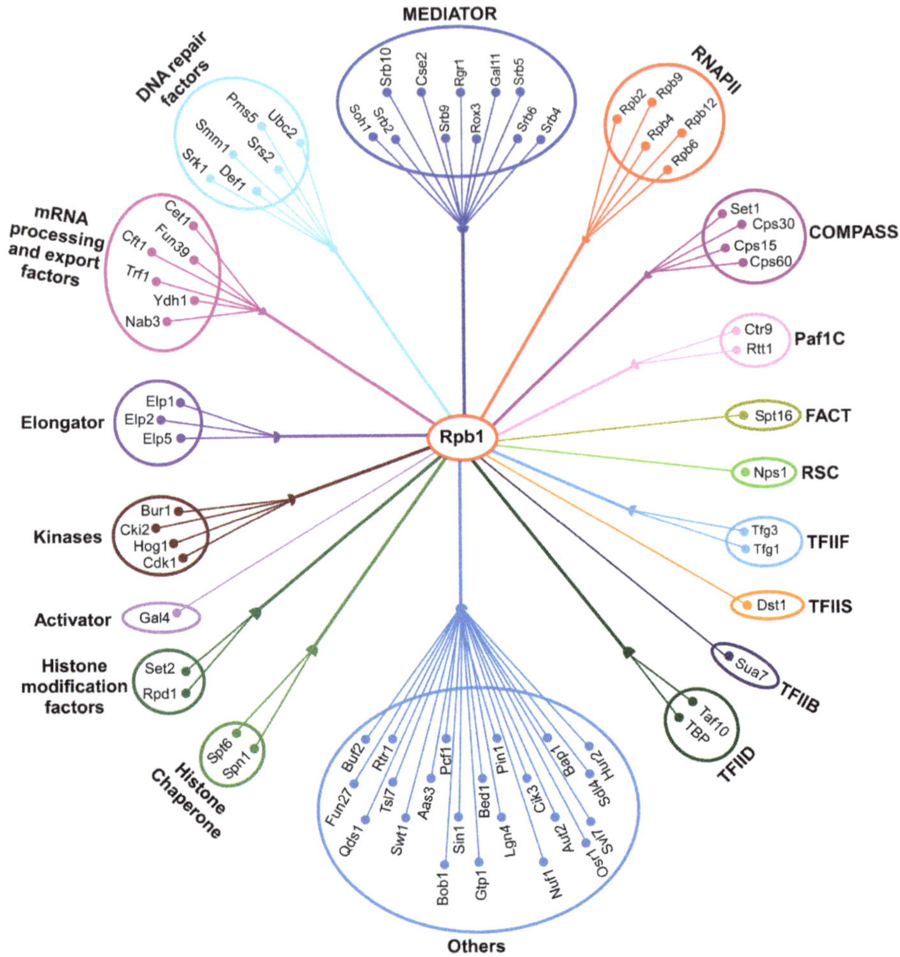

Fig. 1: The interaction of the largest subunit of RNAPII (Rpb1) with other cellular proteins in yeast as revealed by bioinformatic analysis.

stimulatory effect on transcription, whereas the whole Mediator complex represses transcription of a subset of genes (Borggrefe and Kornberg, 2002; Kornberg, 2005; Björklund and Gustafsson, 2005; Casamassimi and Napoli, 2007). Therefore, the Mediator complex plays an important role in transcriptional regulation as a co-activator or co-repressor through its interaction with several transcription factors. One important partner of Mediator in transcription machinery is RNAPII. Here, we have bioinformatically analyzed the interactions of the Mediator components with the largest subunit of RNAPII. This analysis shows that Rpb1 interacts with at least 11 subunits of the Mediator complex. These subunits are Cse2, Srb9, Rgr1, Gal11, Srb4, Srb5, Rox3, Srb2, Srb6, Soh1 and Srb10 as shown in Fig. 1. Likewise, Mediator components interact with other factors during transcription.

Interaction of Rpb1 with COMPASS

COMPASS is a histone methyltransferase in yeast (Miller *et al.*, 2001). It methylates histone H3 on lysine-4 (H3K4) for transcriptional activation (Miller *et al.*, 2001; Santos-Rosa *et al.*, 2002; Liu *et al.*, 2005; Shilatifard, 2006). COMPASS consists of 7 proteins (Cps60, Cps50,

Cps40, Cps35, Cps30, Cps15 and Set1). The Set1 component of COMPASS boasts the enzymatic activity (Miller *et al.*, 2001). Other COMPASS subunits play structural and functional roles to direct Set1 to perform its histone methylase activity (Shilatifard, 2006). COMPASS is recruited to the active genes by RNAPII, and thus methylates histone H3K4 in a transcription-dependent manner (Hampsey and Reinberg, 2003; Shilatifard, 2006). Consistent with these results, this bioinformatic analysis reveals that Rpb1 interacts with four subunits of COMPASS. These subunits are Set1, Cps15, Cps30 and Cps60 (Fig. 1).

Interaction of Rpb1 with PAF

Paf1C is a RNAPII-associated complex. In yeast, Paf1C is composed of Paf1, Cdc73, Ctr9, Rtf1, and Leo1 (Mueller and Jaehning, 2002; Hampsey and Reinberg, 2003). Basically, Paf1C interacts with histone methyltransferase, COMPASS, and histone H2B ubiquitinating enzymes (e.g. Rad6 and Bre1), and thus regulates histone covalent modification for stimulated transcription (Krogan *et al.*, 2003a; Rosonina and Manley, 2005). Paf1C is co-transcriptionally recruited to active genes by RNAPII (Krogan *et al.*, 2003a; Xiao and Strahl, 2005; Rosonina and Manley, 2005). Consistently, this bioinformatic analysis shows the interactions of the Ctr9 and Rtt1 subunits of Paf1C with the largest subunit of RNAPII (Fig. 1). In addition, Paf1C regulates mRNA 3′ end formation through its interaction with mRNA cleavage and polyadenylation factors (Nordick *et al.*, 2008).

Interaction of Rpb1 with FACT

FACT (Facilitates Chromatin Transcription) was identified as a factor required for transcription on chromatin template (Orphanides *et al.*, 1998, 1999). FACT is composed of two subunits, namely, Spt16 and Pob3 in yeast. Lately, FACT has been implicated as an important elongation factor as a histone chaperone (Belotserkovskaya *et al.*, 2003; Belotserkovskaya and Reinberg, 2004). Accordingly, FACT has been shown to travel with elongating RNAPII, and subsequently alters chromatin structure, thereby facilitating transcription (Belotserkovskaya *et al.*, 2003; Belotserkovskaya and Reinberg, 2004; Saunders *et al.*, 2003; Reinberg and Sims, 2006). These results are corroborated with this bioinformatic analysis for the interaction of FACT with the largest subunit of RNAPII (Fig. 1). In addition to its role in transcriptional elongation, FACT is recruited to gene promoter, and regulates transcriptional initiation (Mason and Struhl, 2003). Further, FACT has been shown to participate in regulation of DNA repair and replication (Wittmeyer and Formosa, 1997; Wittmeyer *et al.*, 1999; Schlesinger and Formosa, 2000; Heo *et al.*, 2008). Thus, FACT plays an important role in various cellular events.

Interaction of Rpb1 with RSC

RSC (remodels the structure of chromatin) is an abundant Swi/Snf-like chromatin remodeling complex with multiple subunits in yeast (Cairns *et al.*, 1996). Unlike Swi/Snf, RSC is essential for cellular viability. It has been implicated to activate as well as repress transcription (Moreira and Holmberg, 1999; Sudarsanam and Winston, 2000; Angus-Hill *et al.*, 2001). Further, this bioinformatic analysis shows that the Nps1 component of RSC interacts with the largest subunit of RNAPII (Fig. 1). Thus, RSC appears to regulate transcription via its interaction with RNAPII. In addition to its role in transcription, RSC has been shown to play important role in DNA repair (van Attikum and Gasser, 2005; Chai *et al.*, 2005; Shim *et al.*, 2005).

Interaction of Rpb1 with activator and general transcription factors

In transcriptional activation, the activator stimulates formation of the pre-initiation complex (PIC) assembly which subsequently initiates transcription (Roeder, 2005; Thomas and Chiang, 2006). PIC is formed by the assembly of general transcription factors (e.g. TFIIA, TFIIB, TFIID, TFIIE, TFIIF, TFIIH, RNAPII, etc.). RNAPII is an important component of the PIC assembly, which initiates transcription in the presence of TFIIH. Further, previous studies (Smale and Kadonaga, 2003; Roeder, 2005; Thomas and Chiang, 2006) have implicated the interaction of RNAPII with activator and various general transcription factors during transcriptional initiation or activation as also seen in this bioinformatic analysis (Fig. 1).

Interaction of Rpb1 with Elongator

RNAPII associates with a variety of proteins/protein complexes during transcription elongation. (Shilatifard *et al.*, 2003; Buratowski, 2003; Cramer, 2004; Buratowski, 2005; Shilatifard, 2006; Svejstrup, 2007). Elongator is one such component with six different subunits, namely Elp1, Elp2, Elp3, Elp4, Elp5 and Elp6 (Otero *et al.*, 1999; Krogan and Greenblatt, 2001; Winkler *et al.*, 2001; Li *et al.*, 2001; Svejstrup, 2007). The Elp3 subunit has histone H4 acetyltransferase activity which facilitates transcriptional elongation following RNAPII-dependent association of Elongator with the body of gene (Wittschieben *et al.*, 1999). These results are corroborated with this bioinformatic analysis that reveals that interacts of the Elp1, Elp2 and Elp5 subunits of Elongator with the largest subunit of RNAPII (Fig. 1). In addition to its role in transcriptional elongation, Elongator participates in cytoplasmic kinase signaling, exocytosis and tRNA modification (Holmberg *et al.*, 2002; Rahl *et al.*, 2005; Esberg *et al.*, 2006; Jablonowski *et al.*, 2006; Huang *et al.*, 2005).

Interaction of Rpb1 with Set2 and HDAC

Histone covalent modifications have been associated with transcriptional regulation. Elongating RNAPII has been found to be associated with several covalent modification factors such as Set1 (COMPASS) and Elp3 (Elongator) as mentioned above. Further, elongating RNAPII has been implicated to regulate histone H3 K36 methylation at the body of gene by Set2 histone methylase (Li *et al*,, 2002; Krogan *et al.*, 2003b; Hampsey and Reinberg, 2003; Keogh *et al*,, 2005; Carrozza *et al.*, 2005). Set2 associates with the S-2 phosphorylated form of RNAPII that occurs towards the 3′ end of the coding sequence, hence leading to histone H3 K36 methylation. Intriguingly, methylated-K36 on histone H3 leads to the recruitment of repressive Rpd3(S) complex with multiple subunits (e.g., Rpd3, Eaf3, Ume1, Sin3, Sap30, Sds3, Pho23, and Cti6/Rxt1) (Keogh *et al*,, 2005; Carrozza *et al.*, 2005; Joshi and Struhl, 2005; Yang and Seto, 2008; Lee and Shilatifard, 2007). The chromodomain of Eaf3 is responsible for such recruitment (Joshi and Struhl, 2005; Carrozza *et al.*, 2005; Keogh *et al*,, 2005). Rpd3 has a histone deacetylase (HDAC) activity that preferentially deacetylates acetylated-K on histone H4 (Keogh *et al*,, 2005; Carrozza *et al.*, 2005; Joshi and Struhl, 2005). Such deacetylation has been implicated to repress cryptic transcriptional initiation. Thus, RNAPII regulates transcription by a fine tuning of acetylation/methylation marks via its interaction with Set2 and HDAC (Fig. 1).

Interaction of Rpb1 with histone chaperone proteins

Chromatin presents a structural hindrance for the passage of elongating RNAPII. Therefore, molecular chaperones, which alter chromatin organization, become the rate limiting factors during transcriptional elongation. These chaperones are not only required to clear up the path for RNAPII during elongation, but also maintain the normal chromatin organization after transcription is completed. RNAPII has been shown to associate with molecular chaperones like FACT and Spt6 during transcriptional elongation (Krogan *et al.*, 2002). Consistently, this bioinformatic analysis shows that histone chaperones interact with the largest subunit of RNAPII (Fig. 1). Further, this analysis reveals that Rpb1 interacts with Spn1 that is essential for the recruitment of Spt6 (Zhang *et al.*, 2008). Intriguingly, Spt6 has also been shown to repress transcription of several genes (Compagnone and Osley, 1996). In addition, Spt6 has been implicated in mRNA processing, and export (Bucheli and Burtowski, 2005).

Interaction of Rpb1 with mRNA processing and export factors

Once the transcription is initiated, there are several proteins, which play a significant role in the processing of mRNA. Events, both at the 5′ end (such as capping) as well as 3′ end (e.g. mRNA cleavage and polyadenylation) involve several proteins. The recruitment as well as function of these proteins is tightly regulated. These proteins are shown to be co-transcriptionally recruited to the active genes by the elongating form of RNAPII (Zorio and Bentley, 2004; Bird *et al.*, 2004; Buratowski, 2005; Mandel *et al.*, 2008). In addition, RNAPII co-transcriptionally recruits mRNA export factors (Masuda *et al.*, 2005; Buratowski, 2005). Consistently, this bioinformatic analysis shows the interaction of Rpb1 with the proteins involved in mRNA capping (Cet1), 3′ end formation (Ctf1, Fun39, Nab3 and Ydh1) and export (Trf1) (Fig. 1). Thus, RNAPII appears to regulate mRNA processing and export via its interaction with proteins involved in these processes.

Interaction of Rpb1 with DNA repair factors

Genomic DNA is constantly challenged by internal as well as external genotoxic agents. An extremely cytotoxic ramification is the lesion at the coding region of active gene which inhibits the passage of RNAPII, and subsequently leads to transcription coupled repair through its interaction with DNA repair and related factors (Lainé and Egly, 2006) as also revealed in this bioinformatic analysis (Fig. 1).

Interaction of Rpb1 with protein kinases

Biological processes in the cell are always tightly regulated. Phosphorylation/ dephosphorylation of proteins is one of the major ways the cell synchronize important cellular processes. The phosphorylation status of proteins is maintained by several distinct classes of kinases. Therefore, kinases play a critical role in regulation of many cellular events such as replication, transcription, cell cycle, protein degradation, etc. Further, kinase regulates transcription in a RNAPII-dependent manner. For example the largest subunit of RNAPII interacts with a number of protein kinases such as Cki2, Cdk1, Hog1 and Bur1 as shown in Fig. 1 . These kinases are involved in diverse biological processes such as endocytic trafficking (Cki2), cell cycle regulation (Cdk1), osmoregulation (Hog1), and transcription (Bur1) (Hicke *et al.*, 1998; Alepuz *et al.*, 2003; Keogh *et al.*, 2003; Harvey *et al.*, 2005).

Interaction of Rpb1 with other cellular proteins

In addition to the above interactions, Rpb1 is correlated with several other processes through its interaction with various proteins, which are grouped as "others" in Fig. 1. For example, Rpb1 interacts with proteins involved in replication (e.g., Buf2, Bob1 and Sld4), cytoskeleton organization (e.g., Lgn4, Nuf1 and Gtp1), and protein sorting (e.g., Svl7, Bed1 and Qds1). Further, Rpb1 interacts with some transcriptional regulatory factors such as CCR4-NOT (Fun27), proline isomerase (Pin1), Rtr1, Sin1, Aas3 and Bap1.

Concluding remarks

The function of a protein is dictated by its interaction with other cellular proteins, and, thus, understanding of the protein interaction networks is very crucial for protein function annotation. Several bioinformatic tools have been developed based on experimental data to analyze protein-protein interactions. One such bioinformatic tool is *Cytoscape* software. Using this software, we have analyzed the interactions of the largest subunit of RNAPII with other proteins in yeast to test the validity of this tool for cellular protein-protein interactions in general. We find that the largest subunit of RNAPII interacts with a large number of proteins involved in various stages of transcription, chromatin modification and remodeling, DNA repair, and other biological processes, consistent with experimental studies. Thus, our study validates the general applicability of this bioinformatic approach to analyze cellular protein-protein interactions.

References

Alepuz, P.M., de Nadal, E., Zapater, M., Ammerer, G., and Posas, F. (2003) Osmostress-induced transcription by Hot1 depends on a Hog1-mediated recruitment of the RNA Pol II. EMBO J., *22*, 2433–2442.

Angus-Hill, M.L., Schlichter, A., Roberts, D., Erdjument-Bromage, H., Tempst, P., and Cairns, B.R. (2001) A Rsc3/Rsc30 zinc cluster dimer reveals novel roles for the chromatin remodeler RSC in gene expression and cell cycle control. Mol. Cell *7*, 741–751.

Belotserkovskaya, R., Oh, S., Bondarenko, V.A., Orphanides, G., Studitsky, V.M., and Reinberg, D. (2003) FACT facilitates transcription-dependent nucleosome alteration. Science *301*, 1090–1093.

Belotserkovskaya, R., and Reinberg, D. (2004) Facts about FACT and transcript elongation through chromatin. Curr. Opin. Genet. Dev *14*, 139–146.

Bird, G., Zorio, D.A., and Bentley, D.L. (2004) RNA polymerase II carboxy-terminal domain phosphorylation is required for cotranscriptional pre-mRNA splicing and 3′-end formation. Mol. Cell. Biol. *20*, 8963–8969.

Björklund, S., and Gustafsson, C.M. (2005) The yeast Mediator complex and its regulation.Trends Biochem. Sci. *30*, 240–244.

Borggrefe, T., Davis, R., Erdjument-Bromage, H., Tempst, P., and Kornberg, R.D. (2002) A complex of the Srb8, -9, -10, and -11 transcriptional regulatory proteins from yeast. J. Biol. Chem. *277*, 44202–44207.

Bucheli, M.E, and Buratowski, S. (2005) Npl3 is an antagonist of mRNA 3′ end formation by RNA polymerase II. EMBO J. *24*, 2150–2160.

Buratowski, S. (2003) The CTD code. Nat. Struct. Biol. *10*, 679–680

Buratowski, S. (2005) Connections between mRNA 3′ end processing and transcription termination. Curr. Opin. Cell Biol. *17*, 257–261.

Bushnell, D.A., and Kornberg, R.D. (2003) Complete, 12-subunit RNA polymerase II at 4.1-A resolution: implications for the initiation of transcription. Proc. Natl. Acad. Sci. U.S.A. *100*, 6969–6973.

Cairns, B.R., Lorchm Y., Lim Y., Zhangm M., Lacomism L., Erdjument-Bromage, H., Tempst, P., Du, J., Laurent, B., and Kornberg, R.D. (1996) RSC, an essential, abundant chromatin-remodeling complex. Cell *87*, 1249–1260.

Carrozza, M.J., Li, B., Florens, L., Suganuma, T., Swanson, S.K., Lee, K.K., Shia, W.J., Anderson, S., Yates, J., Washburn, M.P., and Workman, J.L. (2005) Histone H3 methylation by Set2 directs deacetylation of coding regions by Rpd3s to suppress spurious intragenic transcription. Cell *123*, 581–592.

H3 methylation by Set2 directs deacetylation of coding regions by Rpd3S to suppress spurious intragenic transcription. Cell *123*, 581–592.

Casamassimi, A., and Napoli, C. (2007) Mediator complexes and eukaryotic transcription regulation: an overview. Biochimie *89*, 1439–1446.

Chai, B., Huang, J., Cairns, B.R., and Laurent, B.C. (2005) Distinct roles for the RSC and Swi/Snf ATP-dependent chromatin remodelers in DNA double-strand break repair. Genes Dev. *19*, 1656–1661.

Cramer, P., Bushnell, D.A., and Kornberg RD. (2001) Structural basis of transcription: RNA polymerase II at 2.8 angstrom resolution. Science *292*, 1863–1876.

Cramer, P. (2004) RNA polymerase II structure: from core to functional complexes. Curr. Opin. Genet. Dev. *14*, 218–226.

Cramer, P., Armache, K.J., Baumli, S., Benkert, S., Brueckner, F., Buchen, C., Damsma, G.E., Dengl, S., Geiger, S.R., Jasiak, A.J., Jawhari, A., Jennebach, S., Kamenski, T., Kettenberger, H., Kuhn. C.D., Lehmann, E., Leike, K., Sydow, J.F., and Vannini, A. (2008) Structure of eukaryotic RNA polymerases. Annu. Rev. Biophys. *37*, 337–352.

Compagnone-Post, P.A., and Osley, M.A. (1996) Mutations in the SPT4, SPT5, and SPT6 genes alter transcription of a subset of histone genes in *Saccharomyces cerevisiae*. Genetics *143*, 1543–1554.

Egloff, S., and Murphy, S. (2008) Cracking the RNA polymerase II CTD code. Trends Genet. *6*, 280–288.

Esberg, A., Huang, B., Johansson, M.J., and Bystrom, A.S. (2006) Elevated levels of two tRNA species bypass the requirement for Elongator complex in transcription and exocytosis. Mol. Cell *24*, 139–148.

Hampsey, M. and Reinberg, D. (2003) Tails of intrigue: phosphorylation of RNA polymerase II mediates histone methylation. Cell *113*, 429–432.

Harvey, S.L., Charlet, A., Haas, W., Gygi, S.P., and Kellogg, D.R. (2005) Cdk1-dependent regulation of the mitotic inhibitor Wee1. Cell *122*, 407–420.

Heo, K., Kim, H., Choi, S.H., Choi, J., Kim, K., Gu, J., Lieber. M.R., Yang, A.S., and An, W. (2008) FACT-mediated exchange of histone variant H2AX regulated by phosphorylation of H2AX and ADP-ribosylation of Spt16. Mol. Cell *30*, 86–97.

Hicke, L., Zanolari, B., and Riezman, H. (1998) Cytoplasmic tail phosphorylation of the alpha-factor receptor is required for its ubiquitination and internalization. J. Cell Biol. *141*, 349–358.

Holmberg, C., Katz, S., Lerdrup, M., Herdegen, T., Jaattela, M., Aronheim, A., and Kallunki, T. (2002) A novel specific role for IkB kinase complex-associated protein in cytosolic stress signaling. J. Biol. Chem. *277*, 31918–31928.

Huang, B., Johansson, M.J., and Bystrom, A.S. (2005) An early step in wobble uridine tRNA modification requires the Elongator complex. RNA *11*, 424–436.

Jablonowski, D., Zink, S., Mehlgarten, C., Daum, G., and Schaffrath, R. (2006) tRNAGlu wobble uridine methylation by Trm9 identifies Elongator's key role for zymocin-induced cell death in yeast. Mol. Microbiol. *59*, 677–688.

Joshi, A.A., and Struhl, K. (2005) Eaf3 chromodomain interaction with methylated H3-K36 links histone deacetylation to Pol II elongation. Mol. Cell *20*, 971–978.

Keogh, M.C., Podolny, V., and Buratowski, S. (2003) Bur1 kinase is required for efficient transcription elongation by RNA polymerase II. Mol. Cell Biol, *19*, 7005–7018.

Keogh, M.C., Kurdistani, S.K., Morris, S.A., Ahn, S.H., Podolny, V., Collins, S.R., Schuldiner, M., Chin, K., Punna, T., Thompson, N.J., Boone, C., Emili, A., Weissman, J.S., Hughes, T.R., Strahl, B.D., Grunstein, M., Greenblatt, J.F., Buratowski, S., and Krogan, N.J. (2005) Cotranscriptional set2 methylation of histone H3 lysine 36 recruits a repressive Rpd3 complex. Cell *123*, 593–605.

Krogan, N.J., and Greenblatt, J.F. (2001) Characterization of a six-subunit holo-elongator complex required for the regulated expression of a group of genes in *Saccharomyces cerevisiae*. Mol. Cell Biol. *21*, 8203–8212.

Krogan, N.J., Kim,M., Ahn, S.H., Zhong, G., Kobor, M.S., Cagney, G., Emili, A., Shilatifard, A., Buratowski, S., and Greenblatt, J.F. (2002)RNA polymerase II elongation factors of *Saccharomyces cerevisiae*: a targeted proteomics approach. Mol. Cell Biol. *22*, 6979–6992.

Krogan, N.J., Dover, J., Wood, A., Schneider, J., Heidt, J., Boateng, M.A., Dean, K., Ryan, O.W., Golshani, A., Johnston, M., Greenblatt, J.F., and Shilatifard, A. (2003a) The Paf1 complex is required for histone H3 methylation by COMPASS and Dot1p: linking transcriptional elongation to histone methylation. Mol. Cell *11*, 721–729.

Krogan, N.J., Kim, M., Tong, A., Golshani, A., Cagney, G., Canadien, V., Richards, D.P., Beattie, B.K., Emili, A., Boone, C., Shilatifard, A., Buratowski, S., and Greenblatt, J. (2003b) Methylation of histone H3 by Set2 in *Saccharomyces cerevisiae* is linked to transcriptional elongation by RNA polymerase II. Mol. Cell Biol. *23*, 4207–4218.

Kornberg, R.D. (2005) Mediator and the mechanism of transcriptional activation. Trends Biochem. Sci. *30*, 235–239.

Lainé, J.P., and Egly, J.M. (2006) When transcription and repair meet: a complex system. Trends Genet. *22*, 430–436.

Lee, J.S., and Shilatifard,A. (2007) A site to remember: H3K36 methylation a mark for histone deacetylation. Mutat. Res. *618*, 130–134.

Li, J., Moazed, D., and Gygi, S.P. (2002) Association of the histone methyltransferase Set2 with RNA polymerase II plays a role in transcription elongation. J. Biol. Chem. *277*, 49383–49388.

Li, Y., Takagi, Y., Jiang, Y., Tokunaga, M., Erdjument-Bromage, H., Tempst, P., and Kornberg, R.D. (2001) A multiprotein complex that interacts with RNA polymerase II elongator. J. Biol. Chem. *276*, 29628–29631.

Liu, C.L., Kaplan, T., Kim, M., Buratowski, S., Schreiber, S.L., Friedman, N., and Rando, O.J. (2005) Single-nucleosome mapping of histone modifications in S. cerevisiae. PLoS Biol. *3*, e328.

Mandel, C.R., Bai, Y., and Tong, L. (2008) Protein factors in pre-mRNA 3′-end processing. Cell Mol. Life Sci. *65*, 1099–1122.

Mason, P.B. and Struhl, K. (2003) The FACT complex travels with elongating RNA polymerase II and is important for the fidelity of transcriptional initiation *in vivo*. Mol. Cell. Biol. *23*, 8323–8333.

Masuda, S., Das, R., and Cheng, H., Hurt E., and Dorman, N., Reed R.(2005) Recruitment of the human TREX complex to mRNA during splicing. Genes Dev. *19*, 1512–1517.

Miller, T., Krogan, N.J., Dover, J., Erdjument-Bromage, H., Tempst, P., Johnston, M., Greenblatt, J.F., and Shilatifard, A. (2001) COMPASS: a complex of proteins associated with a trithorax-related SET domain protein, Proc. Natl. Acad. Sci. U.S.A. *98*, 12902–12907.

Moreira, J.M. and Holmberg, S. (1999) Transcriptional repression of the yeast CHA1 gene requires the chromatin-remodeling complex RSC. EMBO J. *18*, 2836–2844.

Mueller, C.L. and Jaehning, J.A. (2002) Ctr9, Rtf1, and Leo1 are components of the Paf1/RNA polymerase II complex. Mol. Cell Biol. *22*, 1971–1980.

Nordick, K., Hoffman, M.G., Betz, J.L., and Jaehning, J.A. (2008) Direct interactions between the Paf1 complex and a cleavage and polyadenylation factor are revealed by dissociation of Paf1 from RNA polymerase II. Eukaryot. Cell *7*, 1158–1167.

Orphanides, G., LeRoy, G., Chang, C.H., Luse, D.S., and Reinberg, D. (1998) FACT, a factor that facilitates transcript elongation through nucleosomes. Cell *92*, 105–116.

Orphanides, G., Wu, W.H., Lane, W.S., Hampsey, M., and Reinberg, D. (1999) The chromatin-specific transcription elongation factor FACT comprises human SPT16 and SSRP1 proteins. Nature *400*, 284–288.

Otero, G., Fellows, J., Li, Y., de Bizemont, T., Dirac, A.M.G., Gustafsson, C.M., Erdjument-Bromage, H., Tempst, P., and Svejstrup, J.Q. (1999) Elongator, a multisubunit component of a novel RNA polymerase II holoenzyme for transcriptional elongation. Mol. Cell *3*, 109–118.

Pandit, S., Wang, D., and Fu, X.D. (2008) Functional integration of transcriptional and RNA processing machineries. Curr. Opin. Cell Biol. *20*, 260–265.

Rahl, P.B., Chen, C.Z., and Collins, R.N. (2005) Elp1p, the yeast homolog of the FD disease syndrome protein, negatively regulates exocytosis independently of transcriptional elongation. Mol. Cell *17*, 841–853.

Reinberg, D. and Sims, R.J. 3rd. (2006) de FACTo nucleosome dynamics. J. Biol. Chem. *281*, 23297–23301.

Roeder R.G. (2005) Transcriptional regulation and the role of diverse coactivators in animal cells. FEBS Lett. *579*, 909–915.

Rosonina, E. and Manley, J.L. (2005) From transcription to mRNA: PAF provides a new path. Mol. Cell *20*, 167–168.

Santos-Rosa, H., Schneider, R., Bannister, A.J., Sherriff, J., Bernstein, B.E., Emre, N.C., Schreiber, S.L., Mellor, J., and Kouzarides, T. (2002) Active genes are tri-methylated at K4 of histone H3. Nature *419*, 407–411.

Saunders, A., Werner, J., Andrulis, E.D., Nakayama, T., Hirose, S., Reinberg, D., and Lis, J.T. (2003) Tracking FACT and the RNA polymerase II elongation complex through chromatin *in vivo*. Science *301*, 1094–1096.

Schlesinger, M.B, and Formosa, T. (2000) *POB3* is required for both transcription and replication in the yeast *Saccharomyces cerevisiae*. Genetics *155*, 1593–1606.

Shilatifard, A. (2006) Chromatin modifications by methylation and ubiquitination: implications in the regulation of gene expression, Annu Rev Biochem. *75*, 243–269.

Shilatifard, A., Conaway, R.C., and Conaway, J.W. (2003) The RNA polymerase II elongation complex. Annu Rev Biochem. *72*, 693–715.

Shim, E.Y., Ma, J.L., Oum, J.H., Yanez, Y., and Lee, S.E. (2005) The yeast chromatin remodeler RSC complex facilitates end joining repair of DNA double-strand breaks. Mol. Cell Biol. *25*, 3934–3944

Smale, S.T. and Kadonaga, J.T. (2003) The RNA polymerase II core promoter. Annu Rev Biochem. *72*, 449–479.

Sudarsanam, P. and Winston, F. (2000) The Swi/Snf family nucleosome remodeling
complexes and transcriptional control. Trends Genet *16*, 345–351.

Svejstrup, J.Q. (2007) Elongator complex: how many roles does it play? Curr. Opin. Cell Biol. *19*, 331–336

Thomas, M.C. and Chiang, C.M. (2006) The general transcription machinery and general cofactors. Crit. Rev. Biochem. Mol. Biol. 2006 *41*, 105–178.

van Attikum, H. and Gasser, S.M. (2005) ATP-dependent chromatin remodeling and DNA double-strand break repair. Cell Cycle *4*, 1011–1014.

Winkler, G.S., Petrakis, T.G., Ethelberg, S., Tokunaga, M., Erdjument-Bromage, H., Tempst, P., and Svejstrup, J.Q. (2001) RNA polymerase II elongator holoenzyme is composed of two discrete subcomplexes. J. Biol. Chem. *276*, 32743–32749.

Wittmeyer, J. and Formosa, T. (1997) The *Saccharomyces cerevisiae* DNA polymerase α catalytic subunit interacts with Cdc68/Spt16 and with Pob3, a protein similar to an HMG1-like protein. Mol. Cell. Biol. *17*, 4178–4190.

Wittmeyer, J., Joss, L., and Formosa, T. (1999) Spt16 and Pob3 of *Saccharomyces cerevisiae* form an essential, abundant heterodimer that is nuclear, chromatin-associated, and copurifies with DNA polymerase α. Biochemistry *38*, 8961–8971.

Wittschieben, B.O., Otero, G., de Bizemont, T., Fellows, J., Erdjument-Bromage, H., Ohba, R., Li, Y., Allis, C.D., Tempst, P., and Svejstrup, J.Q. (1999) A novel histone acetyltransferase is an integral subunit of elongating RNA polymerase II holoenzyme. Mol. Cell *4*, 123–128.

Xiao, T., Kao, C.F., Krogan, N.J., Sun, Z.W., Greenblatt, J.F., Osley, M.A., and Strahl, B.D. (2005) Histone H2B ubiquitylation is associated with elongating RNA polymerase II. Mol. Cell Biol. *25*, 637–651.

Yang, X.J. and Seto, E. (2008) The Rpd3/Hda1 family of lysine deacetylases: from bacteria and yeast to mice and men. Nat. Rev. Mol. Cell *9*, 206–218.

Zhang, S., Jin, G., Zhang, X.S., and Chen, L. (2007) Discovering functions and revealing mechanisms at molecular level from biological networks. Proteomics *7*, 2856–2869.

Zhang, L., Fletcher, A.G., Cheung, V., Winston, F., and Stargell, L.A. (2008) Spn1 regulates the recruitment of Spt6 and the Swi/Snf complex during transcriptional activation by RNA polymerase II. Mol. Cell Biol. *28*, 1393–403.

Zorio, D.A. and Bentley, D.L. (2004) The link between mRNA processing and transcription: communication works both ways. Exp. Cell Res. *296*, 91–97.

Chapter 9

Locus Interactions Underlie Seed Yield In Soybeans Resistant to *Heterodera glycines*

U.B. Karangula, M.A. Kassem, L. Gupta, Hany A. El-Shemy, and David A. Lightfoot

Abstract

In soybean (*Glycine max* L. Merr.) combining resistance to cyst nematode (SCN; *Heterodera glycines* I.) with high seed yieldremains problematic. Molecular markers linked to quantitative trait loci (QTL) have not provided a solution. Sets of markers describing a collection of favorable alleles (linkats) may assist plant breeders seeking to combine both traits. The objective of this analysis was to identify linkats in genomic regions underlying seed yield and root SCN resistance QTL. Used were groups of cultivars selected from a single recombinant inbred (RIL) population derived from 'Essex' by 'Forrest' (ExF). The yield was measured at four locations. SCN resistance was determined in greenhouse assays. The mean seed yield was used to define 3 groups (each n=30), high, medium and low. SCN resistance formed 2 groups (SCN resistant (n=21) and SCN susceptible (n=69)). Microsatellite markers (213) alleles were compared with seed yield and root SCN (*Hetrodera glycines)* resistance using mean analysis. The number, size and position of potential linkats were determined. Loci, genomic regions and linkats associated with seed yield were identified on linkage group (LG) K and with root resistance to SCN e on LG E, G, and D1b+W. A method to identify co-localized genomic regions is presented.

Introduction

Soybean [*Glycine max* (L.) Merr.] is one of the world's major crops (Singh and Hymowitz, 1999) grown commercially for its oil (16%) and protein content (42%). Seed yield has doubled over the last 30 years largely by selection of superior adapted cultivars (Specht *et al.*, 2001). However, in the same period soybean has proven susceptible to many diseases that reduce yields. Breeders have struggled to combine high seed yield with resistance to soybean cyst nematode (SCN; caused by *Heterodera glycines* I.) due to linkage drag and other unknown factors (Mudge *et al.*, 2001; Yuan *et al.*, 2002; Concibido *et al.*, 2004; Kopisch-Obuch *et al.*, 2005). In contrast high seed yield has been readily combined with resistance to diseases like soybean sudden death syndrome (SDS; Hnetkovsky *et al.*, 1996; Meksem *et al.*, 1999; Prabhu *et al.*, 1999; Iqbal *et al.*, 2001).

In soybean and other plant species, the majority of economically important agronomic traits like seed yield are controlled in a quantitative fashion (Mansur *et al.*, 1993, 1996). Molecular markers have been used extensively to construct genetic and linkage maps (Song *et al.*, 2004; Zhang *et al.*, 2004; Kassem *et al.*, 2006) and to identify quantitative trait loci (QTL) underlying valuable traits (Yuan *et al.*, 2002; Kopisch-Obuch *et al.*, 2005; Kassem *et al.*, 2006, 2007a,b). However, simple selection of loci for SCN resistance combined with high seed yield has not eliminated the negative relationship between the traits when cyst pressure is low.

Possible explanations for the negative relationship between resistance to SCN and seed yield are sought. One possibility is that introgression of the genes for resistance to SCN has disrupted linkats, a collection of favorable alleles that are linked (Demarly, 1979). In SCN susceptible soybean cultivars large regions where recombination rarely occurs have been detected (Lorenzen *et al.*, 1996; Stefaniak *et al.*, 2005). However, the cultivars with the highest seed yield tend to have more than the mean number of recombination events on several linkage groups (C2, L and M). A second possibility is that the genes for resistance to SCN are themselves unfavorable to seed yield so that selection of new linkats at other genomic locations are needed to ameliorate those negative effects (Lark *et al.*, 1995; Yuan *et al.*, 2002). Indeed there is strong evidence that the resistance allele of major gene for resistance to SCN (*rhg1*) is rigidly co-inherited with an allele on linkage group M, because embryo or pollen in-viability results from breaking this association (Webb *et al.*, 1996). Even

Abbreviations: ExF 'Essex' by 'Forrest'. Linkats, a collection of favorable alleles linked in attraction (linkats). RIL, recombinant inbred population. QTL, quantitative trait loci. SCN, soybean cyst nematode.

when those alleles are co-inherited negative effects on of *rhg1* on plant growth and development can be observed (Yuan *et al.*, 2002; Kopisch-Obuch *et al.*, 2005; Ruben *et al.*, 2006). Both the *rhg1* and *Rhg4* loci contain several candidate genes, but both include receptor-like kinases that appear likely to play a role in root development (Afzal and Lightfoot, 2007).

Recent studies on analysis of soybean genomes (Obermayer and Greilhuber, 1999; Brown-Guedira et al., 2000; Grant *et al.*, 2000; Shoemaker *et al.*, 1996, 2002; Alkharouf and Mathews, 2004; Shultz *et al.*, 2006) showed that the genome duplicated twice and regions of conserved homeology remain (Shoemaker *et al.*, 1996; Grant *et al.*, 2000; Blanc and Wolfe, 2004; Choi *et al.*, 2004; Shultz *et al.*, 2007; Schlueter *et al.*, 2007). The major genes for resistance to SCN on linkage groups A2 (*Rhg4*) and G (*rhg1*) both have 2 or more homeologous regions where conserved synteny is observed (Campbell *et al.*, 2007; Afzal and Lightfoot, 2007). Therefore, it is likely that regions separate from the loci underlying seed yield and resistance to SCN can be affecting those trait but would escape detection by normal method of QTL map development and locus based detection of epistasis.

The objective of this study was to identify interactions between linkats and loci underlying seed yield and SCN resistance. The analysis identified significant patterns of association between alleles at 213 microsatellite DNA markers, soybean genome structures and mean yield. Prediction was made of the genome structure capable of combining yield with disease resistance.

Materials and methods

Genetic material

The genotypes used in this study were the ExF96 population described by Lightfoot *et al.*, 2006. The marker data used was the 213 microsatellite markers from Kassem *et al.*, (2006). Markers were arranged in map order. Trait data were as described in Yuan *et al.*, 2002 for seed yield and in Kassem *et al.*, 2006 for resistance to SCN Hg Type 0.

Data classification

The groups were selected by ranking the genotypes for yield and disease resistance analysis. For seed yield two equal groups of 30 genotypes with lowest yield and highest yield were used. For disease resistance two groups were defined female index reported for five plants challenged with 2000 eggs of SCN Hg type 0 (race3; Yuan *et al.*, 2002). The two groups are resistant (FI<10) and susceptible (FI>10) with 21 and 69 genotypes respectively.

Statistical analysis

The ranked genotypes were used to count the number of alleles corresponding to each marker in map order. Allelic chains (multiple As or Bs) and recombination events (A juxtaposed to B) were counted. The means for each genotype corresponding to A, AA, AAA, AAAA, AAAAA, B, BB, BBB, BBBB, BBBBB, BA, AB, values were calculated. The 12 numbers obtained were then subjected to mean analysis to determine whether phenotype groups differed significantly. The number of allelic chains on each of the chromosome of the genomic structure were calculated and compared.

The mean was the sum of all the scores divided by the number of scores. The formula in summation notation is $\mu = \Sigma X / N$, where μ is the population mean and N is the number of scores. Since the scores were from a sample, then the symbol M refers to the mean and N refers to the sample size. The formula for M is the same as the formula for μ. The mean is a good measure of central tendency for roughly symmetric distributions but can be misleading in skewed distributions since it can be greatly influenced by extreme scores. Therefore, distributions were tested for normality.

Results and discussion

Marker allele segregation

The data set used was identical to that reported in Kassem *et al.*, (2006). There were 12, 958 Forrest alleles and 12, 823 Essex alleles at all markers in the data set. That difference was not significant. Further Chi square showed no significant deviations for the expected 1:1 segregation ratio at any of the 240 markers used (heterogeneous scores were excluded). Therefore, segregation was not biased at any locus or across the whole data set. Recombination events showed evidence for non random distribution across linkage groups whether measure directly (not shown) or as linkats of Essex alleles (Fig. 1) or Forrest alleles (Fig. 2). Linkat and recombination frequencies were significantly different in some genotype groups and some linkage groups (see below).

Correlation among traits

SCN resistance and seed yield in non-SCN infested locations were not negatively correlated ($r=0.12$, $p=0.1$; Supplemental Fig. 1). The negative correlation may have been lost due to a low but significant yield loss to SCN in environments considered non-infested. Cysts were present but at less than 10 cyst per 100 cm^3 of rooting soil.

Disease resistance

Linkats were found on linkage groups (LG) A2, E, D1b and G that were significantly different between the SCN resistant and susceptible phenotype groups ($n=21$ and $n=69$) compared to the overall population distribution. There was a significant increase in the number and length of chains of Forrest alleles associated with the SCN susceptible group at LG E ($p=0.0001$) and D1b ($p=0.01$; Supplental Table 1). Conversely, there was a significant increase in the number and length of chains of Essex alleles associated with the SCN susceptible group at LG A2 ($p=0.045$) and G ($p=0.009$) and a significant decrease on LG E ($p=0.0005$) and D1b ($p=0.05$; Supplemental Table 2). Equally recombination events were depressed among the SCN resistant lines at these LGs. Compared to the population as a whole there was a significant increase in the number of recombination event at LG G and concomitant decrease in the numbers and lengths of chains of Forrest alleles associated with the

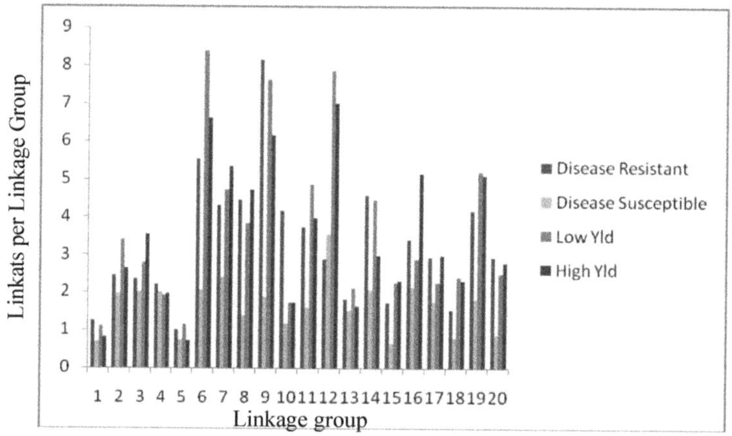

Fig. 1. Linkat distribution of Essex alleles across the 20 linkage groups of soybean within the genotype groups for high yield, low yield, resistance to SCN and susceptibility to SCN. Chromosomes are numbered 1 to 20.

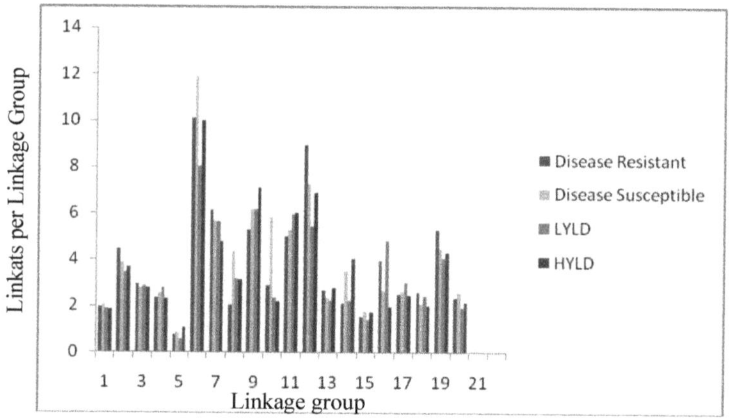

Fig. 2. Linkat distribution of Forrest alleles across the 20 linkage groups of soybean within the genotype groups for high yield, low yield, resistance to SCN and susceptibility to SCN. Chromosomes are numbered 1 to 20.

SCN resistant set. Considering only long linkats, those of more than 3 alleles, there was a significant decrease in the number and length of chains of Essex allele's as among the SCN susceptible lines on LGs A2 (p=0.015), G (p=0.0067), and M (p=0.029). Long linkats of Forrest alleles were found in the disease susceptible group on LG D1b (p=0.012) and LG E (p=0.0002). Overall linkats and recombination events frequencies appeared to be similar in frequency across all LG of the two phenotypic sets but not at the particular linkage groups associated with *Rhg* genes (A2, G and M) present in Forrest and two linkage groups not known to have effects on resistance to SCN HG Type 0 (E and D1b).

Seed yield

Linkats were found on LG I and K that were significantly different between the high yield and low yield phenotype groups (n=30). There was a significant increase (p=0.0098) in the number and length of chains of Essex alleles associated with the high yield group at LG K but a decrease on LG I (p=0.03; Supplemental Table 3). Conversely, LG I showed a significant increase (p=0.002) in the number and length of linkats of Forrest alleles and a significant decrease on LG K (p=0.00003; Supplemental Table 4). Longer linkats of 3 or more alleles showed a similar pattern with a significant increase (p=0.0026) in the number and length of chains of Essex alleles associated with the high yield group at LG K but a decrease on LG I (p=0.018). Also in the high yield group long linkats significantly increased on LG C1 (p=0.026) and significantly decreased LG L (p=0.03) Conversely, LG I showed a significant increase (p=0.002) in the number and length of linkats of Forrest alleles and a significant decrease on LG K (p=0.0003) for both mean and long linkats. Equally recombination events were depressed among both the high and low resistant lines at these LGs. Overall linkats and recombination events frequencies appeared similar in frequency across all LG of the two phenotypic sets except particular LG I and K previously associated with seed yield.

Discussion

Genome architecture associated with yield was detected in the ExF population. Most impressively the markers revealed a very strong linkat to QTL association between yield on genetic linkage group K and I. Both LGs were shown to contain QTL for seed yield (Yuan *et al.*, 2002; Kassem *et al.*, 2007a). In Williams by Essex there were more recombination events and smaller linkats on LG K among high yielding cultivars compared to the population as a whole (Stephaniak *et al.*, 2005). Linkage group I did not differ in linkat size or recombination events in that study. The beneficial allele for the LG K QTL was from Essex (Yuan *et al.*, 2002; Kassem *et al.*, 2006; 2007a) and so the association with long linkats of Essex alleles may indicate a locus with several underlying genes distributed over a large region of the genome. Equally the beneficial allele for the LG I QTL was from Forrest (Yuan *et al.*, 2002; Kassem *et al.*, 2007a) and so the association with long linkats of Forrest alleles may indicate a locus with several underlying genes distributed over a large region of the genome. Several other yield QTL were detected on LG A2, C1, C2, J. and N but they were not associated with linkats or recombination frequency variation (Lark *et al.*, 1995; Kassem etal., 2006; Guzman *et al.*, 2007). Such loci might be underlain by single genes, small gene clusters or sets of epistatic loci.

Genome architecture was associated with SCN resistance but not only at loci known to encompass QTL (LG A2, G and M; Meksem *et al.*, 2001; Webb *et al.*, 1995). Linkats of Forrest alleles detected on linkage groups E and D1b in the SCN susceptible group and their absence from the SCN resistant group might be related to selection for loci that ameliorate the effect of the resistance loci, like that on LG M (Webb *et al.*, 1995). Equally the decrease in the number and length of chains of Essex alleles associated with the SCN resistant group around the QTL bearing regions of LG A2 and G might indicate the loci are multigenenic. However, over the entire genome linkats of Forrest alleles were not more abundant than expected.

The low incidence of polymorphism between Essex and Forrest resulted in a paucity of markers in some regions (C1, J). Gaps between markers in individual maps undoubtedly caused important linkats to be missed.. However, marker density was quite high on most of the LG reported to contain linkats here (A2, C2, G, I, K). With continued expansion of soybean yield QTL studies; additional markers need to be tested in the gaps (Shultz *et al.*, 2007). Further studies can be made by testing the same QTL in a different population and re-examining yield QTL identified without examining the effect of linkats (Orf *et al.*, 1999a; 1999b; Yuan *et al.*, 2002; Kabelka *et al.*, 2004; Smalley *et al.*, 2004; Guzman *et al.*, 2007).

Acknowledgments

The authors acknowledge diverse contributions toward the research reported; for funding from 1991–2004 the ISA, USB, IMBA, NSF, and NCSRB to build the map and trait data; and for assistance by Dr. O. Myers Jr., Dr. M.E. Schmidt and all members of the SIUC field teams from 1983 to 2007. This material was also partly based on work supported by the National Science Foundation under Grant No. 9872635. Any opinions, findings, and conclusions or recommendations expressed in this material are those of the author(s) and do not necessarily reflect the views of the National Science Foundation.

References

Afzal, A.J. and Lightfoot, D.A. (2007) Inclusion bodies contain RHG1 folding intermediates, a novel refolding protocol for protein purification. Prot. Exp. Purif. *53*, 346–355.

Alkharouf, N.W. and Matthews, B.F. (2004) SGMD, the Soybean Genomics and Microarray Database. Nucl. Acids Res. *32*, D398-D400.

Brown-Guedira, G.L., Thompson, J.A., Nelson, R.L., and Warburton, M.L. (2000) Evaluation of genetic diversity of soybean introductions and North American ancestors Using RAPD and SSR markers. Crop Sci. *40*, 815–823.

Blanc, G. and Wolfe, K.H. (2004) Widespread paleopolyploidy in model plant species inferred from age distributions of duplicate genes. Plant Cell *16*, 1667–1678.

Choi, H.K., Mun, J.H., Kim, D.J., Zhu, H., Baek, J.M., Mudge, J., Roe, B., Ellis, N., Doyle, J., Kiss, G.B, Young, N.D., and Cook, D.R. (2004) Estimating genome conservation between crop and model legume species. Proc. Natl. Acad. Sci. U.S.A. *101*, 15289–15294.

Concibido, V.C., Diers, B.W., and Arelli, P.R. (2004) A decade of QTL mapping for cyst nematode resistance in soybean. Crop Sci. *44*, 1121–1131.

Grant, D., Cregan, P.B., and Shoemaker, R.C. (2000) Genome organization in dicots, Genome duplication in Arabidopsis and synteny between soybean and Arabidopsis. Proc. Natl. Acad. Sci. U.S.A. *97*, 4168–4173

Guzman, P.S., Diers, B.W., Neece, D.J., Martin, S.K.S., LeRoy, A.R., Grau, C.R., Hughes, T.J., and Nelson, R.L. (2007) QTL associated with yield in three backcross-derived populations of soybean. Crop Sci. *47*, 111–122.

Hnetkovsky, N., Chang, S.J.C., Doubler, T.W., Gibson, P.T., and Lightfoot, D.A. (1996) Genetic mapping of loci underlying field resistance to soybean sudden death syndrome. Crop Sci. *36*, 393–400.

Kopisch-Obuch, F.J., McBroom, R.L., and Diers, B.W. (2005) Association between SCN resistance loci and yield in soybean. Crop Sci. *45*, 956–965.

Kabelka, E.A., Diers, B.W., Fehr, W.R, LeRoy, A.R., Baianu, I.C., You, T., Neece, D.J., and Nelson, R.L. (2004) Putative alleles for increased yield from soybean plant introductions. Crop Sci. *44*, 784–791.

Kassem, M.A., Shultz, J.L., Meksem, K., Cho, Y., Wood, A.J., Iqbal, M.J., and Lightfoot, D.A. (2006) An updated 'Essex' by 'Forrest' linkage map and first composite interval map of QTL underlying six soybean traits. Theor. Appl. Genet. *113*, 1015–1026.

Kassem, M.A., Meksem, K., Wood, A.J., and Lightfoot, D.A. (2007a) Loci underlying SDS and SCN resistance mapped in the 'Essex' by 'forrest' soybean recombinant inbred lines. Rev. Biol. Biotech. *6*, 2–10.

Kassem, M.A., Meksem, K., Wood, A.J., and Lightfoot, D.A. (2007b) A microsatellite map developed from late maturity germplasm 'Essex' by 'Forrest' detects four QTL for soybean seed yield expected from early maturing germplasm. Rev. Biol. Biotech. *6*, 11–18.

Lark, K.G., Chase, K., Adler, F., Mansur, L.M., and Orf, J.H. (1995) Interactions between quantitative trait loci in soybean in which trait variation at one locus is conditional upon a specific allele at another. Proc. Natl. Acad. Sci. U.S.A. *92*, 4656–4660.

Lightfoot, D.A., Njiti, V.N., Gibson, P.T., Kassem, M.A., Iqbal, M.J., and Meksem, K. (2005) Registration of the Essex by Forrest recombinant inbred line mapping population. Crop Sci. *45*, 1678–1681.

Lorenzen, L.L., and Shoemaker, R.C. (1996) Genetic relationship within old, U.S. soybean cultivar groups. Crop Sci. *36*, 743–752.

Mansur, L.M., Lark, K.G., Kross, H., and Oliveira, A. (1993) Interval mapping of quantitative trait loci for reproductive, morphological, and seed traits of soybean. Theor. Appl. Genet. *86*, 907–913.

Mansur, L.M., Orf, J.H., Chase, K., Jarvik, T., Cregan, P.B., and Lark, K.G. (1996) Genetic mapping of agronomic trait using recombinant inbred lines of soybean. Crop Sci. *36*, 1327–1336.

Meksem, K., Doubler, T.W., Chancharoenchai, K., Njiti, V.N., Chang, S.J.C., Rao-Arelli, A.P., Cregan, P.E., Gray, L.E., Gibson, P.T., and Lightfoot, D.A. (1999) Clustering among loci underlying soybean resistance to *Fusarium solani*, SDS and SCN in near-isogenic lines. Theor. Appl. Genet. *99*, 1131–1142.

Meksem, K., Pantazopoulos, P., Njiti, V.N., Hyten, D.L., Arelli, P.R., and Lightfoot, D.A. (2001) 'Forrest' resistance to the soybean cyst nematode is bigenic, saturation mapping of the *Rhg*1 and *Rhg*4 loci. Theor. Appl. Genet. *103*, 710–717.

Obermayer, R., and Greilhuber, J. (1999) Genome size in Chinese soybean accessions – stable or variable? Ann. of Bot. *84*, 259–262.

Orf, J.H., Chase, K., Jarvik, T., Mansur, L.M., Cregan, P.B., Adler, F.R., and Lark, K.G. (1999a) Genetics of soybean agronomic traits, I. Comparison of three related recombinant inbred populations. Crop Sci. *39*, 1642–1651.

Orf, J.H., Chase, K., Alder, F.R., Mansur, L.M., and Lark, K.G. (1999b) Genetics of soybean agronomic traits, II. Interaction between yield quantitative trait loci in soybean. Crop Sci. *39*, 1652–1657.

Ruben, E., Aziz, J., Afzal, A.J., Njiti, V.N., Triwitayakorn, K., Iqbal, M.J., Yaegashi, S., Arelli, P.R., Town, C.D., Meksem, K., and Lightfoot, D.A. (2006) Genomic analysis of the 'Peking' *rhg1* locus, candidate genes that underlie soybean resistance to the cyst nematode Molec. Genet. Genom. *276*, 320–330.

Schlueter, J.A., Vasylenko-Sanders, I.F., Deshpande, S., Yi, J., Siegfried, M., Rose, B.A., Schlueter, S.D., Scheffler, B.E., and Shoemaker, R.C. (2007) The FAD2 gene family of soybean, insights into the structural and functional divergence of a paleoploid genome. Crop Sci. *47* (S1), S14-S26.

Shoemaker, R.C., Polzin, K., Labate, J., Specht, J., Brummer, E.C., Olson, T., Young, N.D., Concibido, V., Wilcox, J., Tamulonis, J.P., Kochert, G., and Boerma, H.R. (1996) Genome duplication in soybean *(Glycine* subgenus *soja)*. Genetics *144*, 329–338.

Shoemaker, R.C., Keim, P., Vodkin, L., Retzel, E., Clifton, S.W., Waterston, R., Smoller, D., Coryell, V., Khanna, A., J. Erpelding, X. Gai, V. Brendel, C.R. Schmidt, E.G. Shoop, C.J. Vielweber, M. Schmatz, D. Pape, Y. Bowers, B. Theising, J. Martin, M. Dante, T. Wylie, and Granger, C. (2002) A compilation of soybean ESTs, generation and analysis. Genome *45*, 329–338.

Shultz, J.L., Jayaraman, D., Shopinski, K.L., Iqbal, M.J., Kazi, S., Zobrist, K., Bashir, R., Yaegashi, S., Lavu, N., Afzal, A.J., Yesudas, C.R., Kassem, M.A., Wu, C., Zhang, H.B., Town, C.D., Meksem, K., and Lightfoot, D.A. (2006a) The soybean genome database (SoyGD), A browser for display of duplicated, polyploid, regions and sequence tagged sites on the integrated physical and genetic maps of *Glycine max*. Nucleic Acid Research *34*, D1-D8.

Shultz, J.L., Yesudas, C.R., Yaegashi, S., Afzal, A.J., Kazi, S., and Lightfoot, D.A. (2006b) Three minimum tile paths from bacterial artificial chromosome libraries of the soybean (*Glycine max* cv 'Forrest'), Tools for structural and functional genomics. Plant Methods *2*, 9–18.

Singh, R.J. and Hymowitz, T. (1999) Soybean genetic resources and crop improvement. Genome *42*, 605–616.

Smalley, M.D., Fehr, W.R., Cianzio, S.R., Han, F., Sebastian, S.A., and Streit, L.G. (2004) Quantitative trait loci for soybean seed yield in elite and plant introduction germplasm. Crop Sci. *44*, 436–442.

Song, Q.J., Marek, L.F., Shoemaker, R.C., Lark, K.G., Concibido, V.C., Delannay, X., Specht, J.E., and Cregan, P.B. (2004) A new integrated genetic linkage map of the soybean. Theor. Appl. Genet. *109*, 122–128.

SoyBase (2007) http://soybase.agron.iastate.edu/ (Last accessed 12 December 2007).

Specht, J.E., K. Chase, M. Macrander, G.L. Graef, J. Chung, J.P. Markwell, M. Germann, J.H. Orf, and, K.G. Lark (2001) Soybean response to water, A QTL analysis of drought tolerance. Crop Sci. *41*, 493–509.

Stefaniak, T.R., Hyten, D.L., Pantalone, V.R., Klarer, A., and Pfeiffer, T.R. (2005) Soybean cultivars resulted from more recombination events than unselected lines in the same population. Crop Sci. *46*, 43–51.

Stuber, C.W., Lincoln, S.E., Wolff, D.W, Helentjaris, T., and Lander, E.S. (1992) Identification of genetic factors

contributing to heterosis in a hybrid from two elite maize inbred lines using molecular markers. Genetics *132*, 832–839.
Webb, D.M., Baltazar, B.M., Arelli, P.R., Schupp, J., Clayton, K., Keim, P., and Beavis, W.D. (1995) Genetic mapping of soybean cyst nematode race-3 resistance loci in the soybean PI 437.654. Theor. Appl. Genet. *91*, 574–581.
Yuan, J., Njiti, V.N., Meksem, K., Iqbal, M.J., Triwitayakorn, K., Kassem, M.A., Davis, G.T., Schmidt, M.E., and Lightfoot, D.A. (2002) Quantitative trait loci in two soybean recombinant inbred line populations segregating for yield and disease resistance. Crop Sci. *42*, 271–277.
Zhang, W.K., Wang, Y.J, Luo, G.Z., Zhang, J.S., He, C.Y., Wu, X.L., Gai, J.Y., and Chen, S.Y. (2004) QTL mapping of ten agronomic traits on the soybean (*Glycine max*, L. Merr.) genetic map and their association with EST markers. Theor. Appl. Genet. *108*, 1131–1139.

Supplementary Material

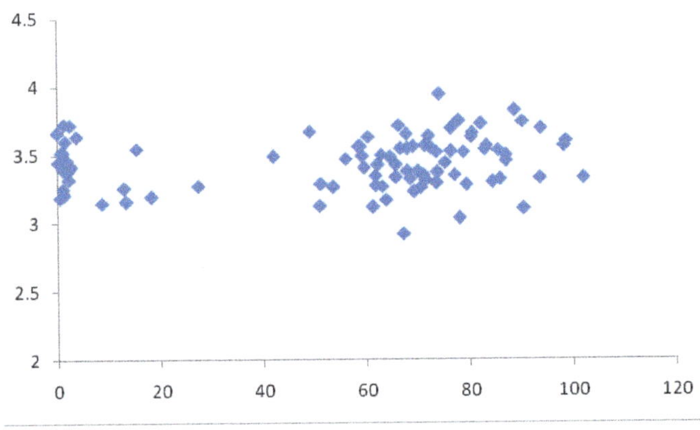

Supplemental Fig. 1. Correlation between mean seed yield in non-infested locations and resistance to SCN Hg Type 0.

Supplemental Table 1. Linkats of Forrest alleles in SCN resistant and susceptible groups.

	Disease Resistant	Disease Susceptible	Ratio	Probability	Linkage Groups
B	2.9048	3	0.96825	0.422409	A1
	6.4286	5.6087	1.1462	0.18244	A2
	5.2857	5.0435	1.048	0.34566	B1
	4.1429	4.3913	0.94342	0.318569	B2
	1.381	1.6957	0.81441	0.151028	***C1***
	12.81	13.957	0.91782	0.256559	***C2***
	8.2857	8.0435	1.0301	0.843459	***D1a***
	3.9524	5.7391	0.68867	0.03494	***D1b***
	7.5238	8.0435	0.93539	0.363385	D2
	4.3333	7.0435	0.61523	0.000105	***E***
	6.619	6.913	0.95747	0.382829	F
	13.19	10.696	1.2333	0.029528	***G***
	3.9524	4.087	0.96707	0.39344	H
	4.1905	5.5217	0.75891	0.044905	I
	2.5238	2.4348	1.0366	0.434052	J
	5.6667	4.5217	1.2532	0.110741	K
	4.381	4.6522	0.9417	0.325899	L
	3.4762	2.913	1.1933	0.157233	M
	7.0952	6.3913	1.1101	0.258952	N
	3.3333	3.8261	0.87121	0.278682	*O*
BB	2	2.0435	0.97872	0.469067	A1
	4.4762	3.8696	1.1568	0.207282	A2
	2.9524	2.7826	1.061	0.399137	B1
	2.381	2.5652	0.92817	0.384325	B2
	0.7619	0.86957	0.87619	0.373186	***C1***
	10.095	11.913	0.84741	0.1666	***C2***
	6.1429	5.6957	1.0785	0.361802	***D1a***
	2.0476	4.3478	0.47095	0.01442	***D1b***
	5.2857	6.1739	0.85614	0.286717	D2
	2.9048	5.8261	0.49858	9.52E-05	***E***
	5	5.3043	0.94262	0.379111	F
	8.9524	7.2609	1.233	0.123539	***G***
	2.6667	2.3913	1.1152	0.324406	H
	2.1429	3.5217	0.60847	0.028879	I
	1.5714	1.7826	0.88153	0.341171	J
	3.9524	2.6957	1.4662	0.093804	K
	2.5238	2.6522	0.9516	0.415432	L
	2.619	2.087	1.255	0.18487	M
	5.2857	4.4783	1.1803	0.23432	N
	2.381	2.6087	0.9127	0.39237	***O***

Supplemental Table 1. Continued

	Disease Resistant	Disease Susceptible	Ratio	Probability	Linkage Groups
BBB	1.4286	1.4783	0.96639	0.461257	A1
	3.1905	2.7391	1.1648	0.265637	A2
	1.4286	1.6087	0.88803	0.372965	B1
	1.5714	1.6087	0.97683	0.473989	B2
	0.57143	0.65217	0.87619	0.390453	***C1***
	8.2857	10.609	0.78103	0.103437	***C2***
	4.7143	4.2174	1.1178	0.332715	***D1a***
	1.4762	3.6087	0.40906	0.012039	***D1b***
	4.0952	5.2609	0.77843	0.227161	D2
	2.2857	5	0.45714	0.000236	***E***
	4	4.1304	0.96842	0.443994	F
	6.8095	5.5217	1.2332	0.175974	***G***
	1.9524	1.7826	1.0952	0.388278	H
	1.381	2.2174	0.62278	0.083346	I
	1	1.3478	0.74194	0.06936	J
	2.9524	1.6522	1.787	0.498311	K
	1.4762	1.4783	0.9986	0.223272	L
	2.0476	1.6087	1.2728	0.363838	M
	3.7619	3.3913	1.1093	0.390396	N
	1.8095	2	0.90476		*O*

Supplemental Table 2. Linkats of Essex alleles in SCN resistant and susceptible groups.

	Disease Resistant	Disease Susceptible	Ratio	Probability	Linkage Groups
A	2.0952	2	1.0476	0.422409	A1
	4.5714	5.3913	0.84793	0.09122	A2
	4.7143	4.9565	0.95113	0.34566	B1
	3.8571	3.6087	1.0688	0.318569	B2
	1.619	1.3043	1.2413	0.151028	***C1***
	8.1905	7.0435	1.1628	0.256559	***C2***
	6.7143	6.9565	0.96518	0.42173	***D1a***
	6.0476	4.2609	1.4193	0.03494	***D1b***
	10.476	9.9565	1.0522	0.363385	D2
	5.6667	2.9565	1.9167	0.000105	***E***
	5.381	5.087	1.0578	0.382829	F
	6.8095	9.3043	0.73186	0.029528	***G***
	3.0476	2.913	1.0462	0.39344	H
	6.8095	5.4783	1.243	0.044905	I
	2.4762	2.5652	0.96529	0.434052	J
	5.3333	6.4783	0.82327	0.110741	K
	4.619	4.3478	1.0624	0.325899	L
	2.5238	3.087	0.81757	0.157233	M
	5.9048	6.6087	0.89348	0.258952	N
	3.6667	4.1739	0.87847	0.27294	***O***

Supplemental Table 2. Continued

	Disease Resistant	Disease Susceptible	Ratio	Probability	Linkage Groups
AA	1.2381	0.69565	1.7798	0.439645	A1
	2.4286	1.9565	1.2413	0.045902	A2
	2.3333	2	1.1667	0.180328	B1
	2.1905	2	1.0952	0.122986	B2
	1	0.73913	1.3529	0.077726	***C1***
	5.5238	2.0435	2.7031	0.371286	***C2***
	4.2857	2.3913	1.7922	0.408344	***D1a***
	4.4286	1.3913	3.183	0.050755	***D1b***
	8.1429	1.8696	4.3555	0.484954	D2
	4.1429	1.1739	3.5291	0.000569	***E***
	3.7143	1.6087	2.3089	0.400358	F
	2.8571	3.5217	0.81129	0.009315	***G***
	1.8095	1.5217	1.1891	0.190319	H
	4.5714	2.0435	2.2371	0.084048	I
	1.7143	0.65217	2.6286	0.361069	J
	3.381	2.1304	1.587	0.165115	K
	2.9048	1.7391	1.6702	0.334952	L
	1.5238	0.78261	1.9471	0.06368	M
	4.1429	1.8261	2.2687	0.279609	N
	2.9048	0.86957	3.3405	0.427126	***O***
AAA	0.71429	0.73913	0.96639	0.472415	A1
	1.1905	2.2174	0.53688	0.015323	A2
	1.3333	1.7826	0.74797	0.247604	B1
	1.2381	0.69565	1.7798	0.064673	B2
	0.52381	0.34783	1.506	0.214849	***C1***
	4.0476	3.7826	1.0701	0.428494	***C2***
	3.2381	3.3043	0.97995	0.475663	***D1a***
	3.2857	2.1304	1.5423	0.07594	***D1b***
	6.5714	6.7391	0.97512	0.452025	D2
	3.381	1.2174	2.7772	0.000449	***E***
	2.619	2.4783	1.0568	0.433379	F
	1.5238	4.0435	0.37686	0.006658	***G***
	1.3333	0.82609	1.614	0.120642	H
	3.0952	2.3478	1.3183	0.16673	I
	1.1429	1.5652	0.73016	0.219074	J
	2.2381	3.1304	0.71495	0.175415	K
	1.9524	1.6957	1.1514	0.34399	L
	0.85714	1.7826	0.48084	0.029304	M
	2.9524	3.3043	0.89348	0.364264	N
	2.381	2.1304	1.1176	0.368596	***O***

Supplemental Table 3. Linkats of Essex Alleles in high and low yield groups.

	Low Yield	High Yield	Ratio	Probability	Linkage Group
A	2.1667	2	1.0834	0.315483	A1
	5.4667	5.4	1.0124	0.067664	A2
	5.1333	5.0667	1.0131	0.37792	B1
	3.5333	4	0.88333	0.138334	B2
	1.6	1.2333	1.2973	0.058823	C1
	10.5	8.6	1.2209	0.299318	C2
	6.9333	7.9333	0.87395	0.102217	D1a
	5.1333	5.4333	0.94478	0.293946	D1b
	10.033	8.8333	1.1358	0.382622	D2
	2.6	2.9667	0.87639	0.448725	E
	6.2667	6.1333	1.0218	0.349282	F
	11.1	9.7667	1.1365	0.376234	G
	3.3333	3	1.1111	0.141052	H
	6.4667	***4.8667***	***1.3288***	***0.054894***	***I***
	3.0667	2.5667	1.1948	0.140526	J
	4.5	***7.2333***	***0.62212***	***1.16E-05***	***K***
	3.9333	4.6333	0.84892	0.065702	L
	2.7667	3.1667	0.87369	0.166634	M
	6.7667	6.9333	0.97597	0.218238	N
	4.9	4.4	1.1136	0.315148	O
	5.31	5.2083	1.02843		Mean
AA	1.0952	0.80952	1.3529	0.221378	A1
	3.381	2.619	1.291	0.10773	A2
	2.7619	3.5238	0.78378	0.1583	B1
	1.9048	1.9524	0.97562	0.468098	B2
	1.1429	0.71429	1.6001	0.108101	C1
	8.381	6.619	1.2662	0.172517	C2
	4.7143	5.3333	0.88394	0.313992	D1a
	3.8095	4.7143	0.80807	0.199123	D1b
	7.619	6.1429	1.2403	0.16879	D2
	1.7143	1.7143	1	0.5	E
	4.8571	3.9524	1.2289	0.215976	F
	7.8571	7	1.1224	0.305542	G
	2.0952	1.619	1.2941	0.221589	H
	4.4286	***2.9524***	***1.5***	***0.032979***	***I***
	2.2381	2.2857	0.97917	0.468454	J
	2.8571	***5.1429***	***0.55554***	***0.009804***	***K***
	2.2381	2.9524	0.75806	0.178496	L
	2.381	2.2857	1.0417	0.436078	M
	5.1905	5.0952	1.0187	0.469325	N
	2.4762	2.7619	0.89656	0.317056	O
	3.6571	3.50952	1.07985		Mean

Supplemental Table 3. Continued

	Low Yield	High Yield	Ratio	Probability	Linkage Group
AAA	0.83333	0.6	1.3889	0.227018	A1
	2.1667	2.3	0.94204	0.395092	A2
	1.9333	2	0.96665	0.460807	B1
	1.1	1.0667	1.0312	0.465053	B2
	0.73333	**0.3**	**2.4444**	**0.026908**	C1
	5.7667	5.0667	1.1382	0.304805	C2
	3.5	4.2333	0.82678	0.211233	D1a
	2.5	3.1333	0.79788	0.209991	D1b
	6.6333	5.4	1.2284	0.14822	D2
	1.3	1.6	0.8125	0.253408	E
	3.4333	**3.4333**	**1**	**0.5**	F
	5.4667	4.8333	1.131	0.309594	G
	1.5667	1.2	1.3056	0.207928	H
	3.1	**1.8333**	**1.6909**	**0.018296**	***I***
	1.7333	1.5667	1.1063	0.359631	J
	1.6667	**3.9333**	**0.42374**	**0.002573**	***K***
	1.0333	**2.0667**	**0.49998**	**0.028137**	L
	1.4333	1.5333	0.93478	0.407889	M
	3.5667	4.4	0.81061	0.186029	N
	2.6333	2.4667	1.0675	0.396604	O
	2.68999	2.6483			Mean

Supplemental Table 4. Linkats of Forrest alleles in high and low yield groups.

	Low Yield	High Yield	Ratio	Probability	Linkage Group
B	2.8333	3	0.94443	0.339754	A1
	5.5333	5.6	0.98809	0.447336	A2
	4.8667	4.9333	0.9865	0.455683	B1
	4.4667	4	1.1167	0.178333	B2
	1.4	1.7667	0.79244	0.084969	C1
	10.5	12.4	0.84677	0.100339	C2
	8.0667	7.0667	1.1415	0.148827	D1a
	4.8667	4.5667	1.0657	0.357885	D1b
	7.9667	9.1667	0.86909	0.163227	D2
	3.4	3.0333	1.1209	0.232472	E
	7.7333	7.8667	0.98304	0.450061	F
	8.9	10.233	0.86974	0.165103	G
	3.6667	4	0.91668	0.238494	H
	4.5333	6.1333	0.73913	0.006745	***I***
	1.9333	2.4333	0.79452	0.146559	J
	6.5	3.7667	1.7256	9.48E-05	***K***
	5.0667	4.3667	1.1603	0.090635	L
	3.2333	2.8333	1.1412	0.19872	M
	6.2333	5.9667	1.0447	0.393425	N
	3.1	3.4333	0.90292	0.328974	O
SEM	2.435599	2.796534	0.212329		
SUM	104.8	106.5664	1.007498	MEAN	

Supplemental Table 4. Continued

	Low Yield	High Yield	Ratio	Probability	Linkage Group
BB	1.9	1.8667	1.0178	0.471703	
	3.4667	3.7	0.93695	0.340583	
	2.8667	2.8	1.0238	0.45589	
	2.8	2.3333	1.2	0.218565	
	0.6	1.1	0.54545	0.035078	
	8.0333	10	0.80333	0.095882	
	5.6333	4.7667	1.1818	0.188766	
	3.2	3.1333	1.0213	0.468363	
	6.2	7.1	0.87324	0.238981	
	2.3667	2.2333	1.0597	0.411216	
	5.9667	6.0333	0.98896	0.473981	
	5.4333	6.9	0.78743	0.133041	
	2.2667	2.8	0.80954	0.146998	
	2.2667	4.0333	0.562	0.00263	
	1.4333	1.7333	0.82692	0.244333	
	4.8333	1.9667	2.4576	2.98E-05	
	3.0333	2.5	1.2133	0.154261	
	2.4333	2.0333	1.1967	0.206271	
	4.1	4.3	0.95349	0.418899	
	1.9333	2.1667	0.89228	0.370876	
SEM	1.901279	2.271715	0.388184		
SUM	70.7666	73.4999	1.01758	MEAN	
BBB	1.2667	1.2	1.0556	0.430955	
	2.4667	2.5667	0.96104	0.426163	
	1.8333	1.5	1.2222	0.257691	
	2	1.6	1.25	0.224322	
	0.3	0.76667	0.3913	0.018609	
	6.5	8.3333	0.78	0.105269	
	4.1667	3.4667	1.2019	0.211313	
	2.3333	2.3667	0.98589	0.482938	
	5	5.9333	0.8427	0.22218	
	1.9667	2	0.98335	0.478212	
	4.5667	4.4	1.0379	0.42911	
	3.7	5.5333	0.66868	0.060731	
	1.5667	1.9667	0.79661	0.213927	
	1.1667	2.7667	0.42169	0.002201	
	1.0333	1.1667	0.88566	0.367227	
	3.5333	1.2	2.9444	0.000314	
	1.7333	1.3667	1.2682	0.202926	
	1.9667	1.4667	1.3409	0.146063	
	2.8	3.1333	0.89363	0.350178	
	1.5333	1.6333	0.93877	0.5	
SEM	1.956247	0.51542	0.168429		
MEAN			1.043521		

Chapter 10

Different Responses of Two Genes Associated with Disease Resistance Loci in Maize (*Zea mays* L.) to 3-allyloxy-1,2-benzothiazole 1,1-dioxide

Jiazheng Yuan, Jennifer Tedman, Liakat Ali, Jie Liu, Jeff Taylor, David A. Lightfoot, Michiaki Iwata, and K. Peter Pauls

Abstract

Probenazole (3-allyloxy-1,2-benzothiazole 1,1-dioxide, PBZ) is a bactericide and fungicide that acts by inducing plant defense systems. It has been shown to induce the expression of NBS-LRR genes like *RPR1* (rice probenazole-response gene) in rice (*Oryza sativa* L.) and systemic acquired resistance (SAR)-like disease resistance. Two maize (*Zea mays* L.) genes *Zmnbslrr1* (a NBS-LRR gene, cloned from a disease resistance analog PIC11 based) and *Zmgc1*, (a putative guanylyl cyclase-like gene) have both been associated with quantitative resistance loci (QTL) for resistance to *Fusarium graminearum*. PIC11 was associated with Fusarium stalk rot and ZmGC1 showed resistance to Gibberella ear rot caused by *F. graminearum*. The objectives of the current study here were to characterize the *Zmnbslrr1* gene and to determine whether it and *Zmgc1* respond to the inducer PBZ. The transcript abundance of *Zmnbslrr1* expression was significantly reduced in corn seedlings of the Gibberella ear rot resistant genotype CO387 48 h after PBZ treatment. In contrast, the transcript abundance of the maize *Zmgc1* gene increased more than 10-fold 8h after the treatment. Therefore, the two genes do not appear to be coordinately regulated by PBZ.

Introduction

Plants utilize chemical barriers, consisting of antimicrobial compounds, physical barriers such as reinforced cell walls, and sensing barriers, including arrays of resistance (R) genes, to prevent pathogen infection (Dangl and Jones, 2001). Pathogen infestation can cause accumulations or reductions in transcript levels (Iqbal *et al.*, 2005). Plant disease resistance and response genes can be classified into many distinct groups based on the characteristics of the proteins that they encode and the regulons that control their expression. The largest group of such genes encodes proteins with a nucleotide binding site (NBS) motif in the amino-terminal domain followed by a variable number of leucine rich repeats (LRRs) domain in carboxy-terminal domain (Meyers *et al.*, 2003; Belkhadir *et al.*, 2004). The NBS domain may function in ATP hydrolysis and signal transduction while the LRR is hypothesized to be responsible for recognizing a pathogen derived signal (Belkhadir *et al.*, 2004). NBS-LRR genes are wide-spread in plants and may also be involved in plant development and responses to abiotic stresses (Michelmore, 2000).

In some situations, a single resistance gene might provide protection against one or more strains of a particular pathogen when introduced into a susceptible plant of the same species. However, many disease resistance genes recognize only very limited number of pathogen isolates and therefore, they are the quickly defeated by co-evolving pathogens (Pink, 2002). The evolutionary tug of war, often called the "Red Queen Hypothesis" assumes that plant pathogens will constantly evolve to overcome plant defense barriers (Clay and Kover, 1996).

For a more sustainable plant pathogen management system, partial resistance combined with chemical control may be an option. Probenazole (3-allyloxy-1,2-benzisothiazole 1,1-oxide; PBZ) has been widely used with partially resistant rice (*Oryza sativa*) cultivars to control rice blast disease for decades in Asia (Watanabe *et al.*, 1979). PBZ induces systemic acquired resistance (SAR)-like disease resistance in rice (Sakamoto *et al.*, 1999). PBZ was registered for use in several other crops mainly against bacterial pathogens (Oostendorp *et al.*, 2001). The main metabolites of probenazole in rice are 2-sulfamoylbenzoate, saccharin, and N-D-glucopyranosylsaccharin (Uchiyama *et al.*, 1973). Part of the effect of PBZ against pathogens may be due to weak fungitoxicity against the rice pathogen *Pyricularia oryzae*.

However, similarities between probenazole induced SAR-like resistance and the SAR-like resistance induced by pathogen suggested that the protective effects of the chemical were mainly due to activation of host defense mechanism. Treated plants respond more quickly to infection (Langcake *et al.*, 1983). Sakamoto *et al.*, (1999) identified a gene, named the *RPR1* (rice probenazole responsive) gene, which was induced by treatment with probenazole, other chemical inducers of systemic acquired resistance (SAR) such as benzothiadiazole (BTH) and salicylic acid (SA), as well as pathogens. The RPR1 protein contained NBS and LRR domains similar to several known R gene products such as RPM1 (Grant *et al.*, 1995), Tobacco N (Whitham *et al.*, 1994), and Mi (Rossi *et al.*, 1998). PR1 has antifungal activities and its synthesis serves as a biomarker of plant responses to pathogens (Durner *et al.*, 1998; Ryals *et al.*, 1996). In addition, enhanced activity of cell membrane GTPase (EC number: 3.6.5.5), an enzyme involved in intracellular signal transduction pathways, has been observed in rice plants treated with probenazole (Kanoh *et al.*, 1993).

Guanylyl cyclases (EC 4.6.1.2 also called guanylate cyclases) catalyze the formation of guanosine 3′,5′-cycline monophosphate (cGMP) from guanosine 5′-triphosphate (GTP; Bowler *et al.*, 1994). cGMP is a ubiquitous second messenger in plants (Bowler *et al.*, 1994) and Durner *et al.*, (1998) demonstrated that it functions as a second messenger in the induction of PAL (phenylalanine ammonia lyase) by exogenous nitric oxide (NO). Therefore, cGMP was associated with the synthesis of phenylpropanoid products and defense functions. In addition, reactive oxygen intermediates (ROI) induce hypersensitive cell death by raising cytosolic calcium (Levine *et al.*, 1996) and NO can reduce the threshold for calcium signaling by modulating cGMP-gated ion channels (Berridge, 1993).

Fusarium graminearum, Schwabe infects a wide variety of cereal grains, including: rye, barley, wheat, oats, and maize. The pathogen causes Gibberella ear rot and Fusarium stalk rot in maize and wheat and barley head blight in temperate areas (Desjardins, 2003). Gibberella ear rot is characterized by the growth of reddish-white mold beginning at the tip of the ear that can cover the entire ear (Sutton, 1982). Infected grain contains mycotoxins, such as deoxynivalenol (DON, vomitoxin) and zearalenone (ZEN) (Greenhalgh *et al.*, 1983; Molto *et al.*, 1997). Consumption of contaminated food is dangerous to livestock and humans. Mycotoxin contaminated grain fed to livestock (especially swine) induces emesis characterized by vomiting, feed refusal, and decreased weight gain (Christensen and Meronuck, 1986).

Resistance to Gibberella ear rot in maize is quantitative (partial), specific to the mode of fungal entry (silk channels or kernel wounds) and is highly influenced by the environment (Sutton, 1982; Reid *et al.*, 1996; Chungu *et al.*, 1996; Ali *et al.*, 2005). Partial resistance to the related *Fusarium* pathogens of legumes, including *Nectria hematococca* and *Fusarium solani* f. sp. *glycines* (now *Fusarium virguliforme* Roy [Aoki]), has been shown to prevent the loss of transcripts from genes like PAL, CHS, and PR1 in their host species during infection (Covert *et al.*, 2001; Iqbal *et al.*, 2005). The molecular mechanism of Gibberella ear rot resistance in maize is still not well characterized.

Maize NBS-LRR resistance gene analogue (RGA) PIC11 (Collins *et al.*, 1998) was mapped to chromosome 5. It was located a genomic region that defined a QTL associated with corn Gibberella stalk rot resistance (Pe *et al.*, 1993). A maize gene (*Zmgc1*) encoding a putative 267-amino acid guanylyl cyclase-like protein (ZmGC) was shown to be significantly associated with Gibberella ear rot across different environments. (Yuan *et al.*, 2008). The *Zmgc1* coding sequences were identical in a Gibberella ear rot-resistant line (CO387) and a susceptible line (CG62) but several differences in nucleotide sequences were identified in their UTRs and introns. Infection with *F. graminearum* increased the transcript abundance of *Zmgc1* in both genotypes within hours (Yuan *et al.*, 2008).

A 1674 bp genomic DNA fragment encoding an open reading frame (ORF, *Zmnbslrr1*) was isolated in our current work based on the sequence information of the RGA PIC11 to investigate the role of this gene in Gibberrella ear rot resistance and to test its inducibility by PBZ. The Blast results indicate that the ORF is an NBS-LRR gene shared the most significant similarity with genes that possess motifs of nucleotide binding sites (NBS) and leucine-rich repeats (LRR) including RPR1 (protein induced in rice treated with probenazole) (Sakamoto *et al.*, 1999), LRR19 (Madsen *et al.*, 2003), and RXO1 (Zhao *et al.*, 2005). The overall objective of the current study was to determine whether probenazole treatment could affect the activity of two genes that were associated with resistance to *F. graminearum*. We found that the *Zmnbslrr1* and *Zmgc1* genes responded to probenazole differently, suggesting that the two genes were not regulated in the same way by PBZ.

Materials and methods

Plant materials

Maize seeds were from a F_5 recombinant inbred (RI) population (n = 144) was derived by single seed descent from a cross between the inbred lines, CG62 and CO387 (Ali *et al.*, 2005). CG62 was developed at the University of Guelph from Pioneer 3902 (Lee *et al.*, 2001) and is susceptible to *F. graminearum* after silk or kernel inoculation (Ali *et al.*, 2005). CO387, developed at Agriculture and Agri-Food Canada (AAFC, Ottawa, ON), has resistance to Gibberella ear rot following silk or kernel inoculation (Reid *et al.*, 1994; Reid and Hamilton, 1996). This F_5 population was previously genotyped with 162 molecular markers and phenotyped for Gibberella ear rot disease in 4 environments, resulting in the identification of 18 QTL for resistance to *F. graminearum* (Ali *et al.*, 2005).

Nucleic acid isolation

DNA for cloning the *Zmnbslrr1* gene and gene mapping was obtained from maize (*Zea mays* L.) leaf samples. The leaves were harvested from field-grown plants (F_5), freeze-dried, and ground to fine powder. DNA was extracted from the powdered leaf material using the CTAB (Cetyltrimethyl ammonium bromide) method (Hoisington *et al.*, 1994). The DNA was treated with RNase and stored in Tris-EDTA buffer.

To collect RNA samples, seeds of inbred CO387 were germinated in a commercial soil mixture (Terra-lite Redi-earth, WR Grace & Co., Canada Ltd., Ajax, ON). The pots (30 x 30 x 5cm) were maintained in a greenhouse kept at 24°C with 800 microeinsteins/m^2/sec of light on a 16-h photoperiod. Two-week-old seedlings were arranged in a completely randomized design and sprayed with 2% (w/w) probenazole (Meijiseika Co., Tokyo, Japan) to foliar runoff. At intervals starting before the chemical spray and ending 6 days after the treatment, plants were harvested to obtain leaf samples. Three plants were harvested for the treatment at each time points. The plant tissues were immediately frozen in liquid nitrogen and kept at –80°C. Total RNA was isolated using the RNeasy plant mini kit (Qiagen, Mississauga, ON) in combination with DNase treatment using RNase-free DNase kit (Qiagen GmbH, Hilden, Germany), according to the manufacturer's instructions.

Genome walking and sequence analyses

Primers were synthesized by Sigma-Aldrich (Sigma-Aldrich, Mississauga, ON). The universal GenomeWalker™ kit (BD Biosciences, Palo Alto, CA) was used to obtain sequences adjacent to the previously characterized 420 bp PIC11 (RGA) fragment (Collins *et al.*, 1998). In short, DNA libraries were constructed by digesting genomic DNA with four different restriction enzymes (in 100 µl mixtures containing 3 µg of genomic DNA, 80 units of restriction enzyme and 1× restriction buffer) and adaptors were ligated to the ends of the fragments according to manufacturer's instructions. Several GC rich (>40%) gene specific primers (gsps, 26–30 bp) were designed for Genome-walking on both directions based on sequence information for the 420 bp PIC11 RGA fragment (Collins *et al.*, 1998). The outer adaptor primer (AP1) provided in the kit, a gene-specific primer gsp1 (5′-TGACATGCTTGCTCTCTGATCTATTAGCT-3′) and DNA of each of the libraries were used for the primary PCR reaction. A single, major PCR product from at least one of the four libraries was gel purified using a gel purification kit (Qiagen, Mississauga, ON), cloned into TOPO TA using the TOPO cloning kit (Invitrogen, Burlington, ON) and introduced into *Escherichia coli* One-Shot competent cells according to the manufacturer's instructions (Invitrogen, Burlington, ON). Several clones were picked and plasmid DNA was isolated with a Qiagen Miniprep Kit (Qiagen, Mississauga, ON). Cloned fragments were sequenced with a CEQ2000 sequencer (Beckman-Coulter, Fullerton, CA) after dideoxy nucleotide sequencing reaction using DTCS Quick Start sequencing kit (Beckman-Coulter, Fullerton, CA). All of the DNA fragments began in a known sequence in GSP1 and extended into the unknown adjacent genomic DNA. The new sequence information was used to design new gene specific primers for further walking.

Primers PIC11-F 5′-ATTCAGACATGGCA GATGCCCTATTTGTA-3′, and PIC11-R 5′-GTCCC TCTACTTACACCGCGAACTACCTA-3′ were used to obtain a PCR fragment encompassing the complete open reading frame of the *Nbslrr1* gene from CO387 genomic DNA after three steps of GenomeWalking. The final amounts in the PCR mixture used to clone the entire gene were: 10–20 ng template, 0.4 M of each primer, 1× PCR buffer, 2 mM $MgSO_4$, 0.2 mM dNTP and 1 unit high fidelity Platinum *Taq* (Invitrogen, Burlington, ON). The reaction was performed under the following conditions: denaturation at 94°C for 3 min, 35 cycles with denaturation at 94°C for 30 s, annealing at 55–65°C for 45 s, extension at 68°C for 3–6 min and a final hold at 10°C. The PCR samples were electrophoresed through 0.8 % (w/v) agarose gels, stained with 0.5 µg/ml ethidium bromide and visualized under UV light. The DNA fragment was cut from the gel, purified, cloned, and sequenced as described in the preceding section.

RFLP and SCAR analyses

Genomic DNA samples (20 µg) from lines in the population were digested with restriction enzymes (30 units of *Eco*RV), and fractionated by electrophoresis through 0.8 % agarose gels. The gels were stained with ethidium bromide, visualized under UV light and the DNA was transferred onto nylon membranes (Hybond-N+; Amersham Biosciences, Arlington Heights, IL) using protocols recommended by the manufacturer. The forward primer (RFLP-PIC11-F 5′-CTAATAGATCAGAGAGCAA GCA-3′) and the reverse primer (RFLP-PIC11-R 5′-GTTCCCAGGATTCAGCATATTG-3′) for the 420 bp PIC11 fragment were used with genomic DNA to PCR amplify a probe for hybridization. The RFLP probe was labeled Digoxigenin-11-dUTP into the PCR reaction. The membranes were pre-hybridized overnight at 37–42°C in DIG Easy-Hyb solution and hybridized in DIG Easy Hyb solution containing the PIC11 probe for at least 24 hours at 37–42°C according to the instructions of manufacturer (Boehringer Mannheim, Germany). After several washings with 2X SSC in room temperature and 0.5X SSC at 65°C, the hybridization pattern was visualized by chemiluminescence (with CDP-star as the substrate, Roche Diagnostics, Laval, PQ) and recorded by exposure of X-ray film (Kodak Inc., Rochester, NY).

A SCAR (sequence-characterized amplified region) marker, developed from the total *Zmnbslrr1* sequence fragment, was used to screen entire mapping population (data not shown). The 2100 bp fragment was PCR amplified using primers SCAR-PIC11-F 5′- ACTGGGCATGAAGGGTACTAG-3′ and SCAR-PIC11-R 5′- CTTAATCAGGTCCCTGGTTAGT-3′. The PCR conditions were as following: denaturation at 94°C for 3 min, 35 cycles with denaturation at 94°C for 30 s, annealing at 60–65°C for 45 s, extension at 72°C for 3 min and final hold at 10°C. Following amplification, the samples were electrophoresed through 0.8 % (w/v) agarose gels, stained with 0.5 µg/ml ethidium bromide and visualized under UV light.

Mapping population and statistic analysis

PIC11 (Zmnbslrr1 anchored) was mapped by adding the RFLP and SCAR F5 segregation data was to the previous scores for RAPD, RFLP, SSR and SCAR markers for the individuals in the CG62 x CO387 RIL population (Ali *et al.*, 2005) and reanalyzing linkages with MAPMAKER/EXP version 3.0b (Lander *et al.*, 1987) with the Kosambi mapping function (Kosambi, 1944). Linkage groups were identified using the "Group" command with a LOD score of 4.0 – 6.0 and a recombination fraction of 0.4 (Ali *et al.*, 2005). The *Zmnbslrr1* (PIC11) derived RFLP markers were added to existing linkage groups with the "Try" command. To determine the locations of the QTL on the map and estimate their phenotypic effects, composite interval mapping (CIM) was performed using QTL

Cartographer V2.0 (Wang *et al.*, 2006; Ali *et al.*, 2005). The chromosome orientations followed the Intermated B73 x Mo17 (IBM) maize map (Davis *et al.*, 1999; http://www.maizegdb.org) with the lowest bin positions at the top. The PIC11-RFLP and SCAR marker were evaluated individually by χ^2 analyses for goodness of fit against a 1:1 segregation ratio at a 0.01 probability level. The RFLP and SCAR associations with *F. graminearum* resistance were tested by a single marker analysis using an ANOVA (V8.2, SAS Institute, Cary, NC, USA).

BLAST searches were performed with default settings using the National Center for Biotechnology (NCBI) search engine (http://www.ncbi.nlm.nih.gov/BLAST). Sequence alignments and phylogenetic analyses were performed by using Clustal W (Thompson *et al.*, 1994) for nucleotide and Clustal X (Thompson *et al.*, 1997) for amino acid with default settings at site of the European Bioinformatics Institute (http://www.ebi.ac.uk/clustalw).

Real-time PCR quantification

A Ribogreen RNA quantitation kit (Molecular Probes, OR) and a Wallace Victor microplate reader (Perkin-Elmer Life Sciences, St. Laurent, PQ) with the lamp filter F485 nm and the emission filter F530 nm were used to quantity the RNA. RNA quality (ribosomal band integrity) was determined by separating 2 µg of total RNA on 1× TBE of 0.8% (w/v) agarose gel and staining with 0.5 µg/ml ethidium bromide. Reverse transcription was carried out using 1 µg of total RNA for all samples with the Restroscript kit at 50°C for one hour and 92°C for 5 min, following manufacturer's instruction (Ambion, Austin, TX).

Quantitative real-time PCR was carried out with an iClycler (BIO-RAD Laboratories, Mississauga, ON) with the Real-time Optical Module (BIO-RAD Laboratories, Mississauga, ON) and a SYBR Green Supermix kit (BIO-RAD). The real-time PCR primers that were used were: Zmgc1-Realtime-F 5′-ACAACAGCACAACG ATGAGCCTGACTAC-3′ and Zmgc1-Realtime-R 5′-ATTTGCGGGCCTCGTCCAAGCTCTTCATTG-3′, which were based the sequences obtained for exon 6 and exon 8 of the *Zmgc1* gene, respectively, and Zmnbslrr1-realtime-F 5′-AGCTAATAGAT CAGAGAGCAAGCATGTCA-3′ and Zmnbslrr1-realtime-R 5′-TCCTTTCTCCGAGTAGTTATCAGCACTTTG-3′. Both real time PCR products were approximately 150 bp. The final 20 µl PCR reaction mixture contained: 2 µl of cDNA, 4 nM primers and 10 µl 2X SYBR green supermix (BIO-RAD, Hercules, CA). Gene copy number was calculated using a standard curve method with regression equations generated by iCyclerIQ™ the on the iCycler (BIO-RAD Laboratories, Mississauga, ON). The internal standard used to normalize the results was a fragment amplified with primers derived from the maize 18S rRNA (F: 5′-TAGTTAGCAGGCTGAGGTCTCG;-3′; R: 5′-CTTACCAGGTCCAGACATAGC-3′). A melting curve analysis was performed to distinguish non-specific amplification from target amplicons by heating reaction from 50°C to 95°C at a rate of 0.5°C/s. All three biological samples obtained at each time interval were run in triplicate, as were samples for generating the standard curve and the no template controls. Analysis of variance (ANOVA) with Tukey's HSD test was employed to determine the differences between treatment intervals.

Results

Gene structure of Zmnbslrr1 and Zmgc1 gene

In total 2300 bp of genomic DNA from the downstream and upstream of the 420 bp PIC11 RGA fragment (Collins *et al.*, 1998) was cloned and sequenced by performing genome walking steps with genomic DNA from genotype CO387 (data not shown). A 1674 bp open reading frame (ORF, *Zmnbslrr1*) was identified in the sequence (Fig. 1). The region upstream of the gene was GC-rich and appears to be a TATA-less promoter (Smale, 1997). The ORF was not interrupted by introns. *Eco*RV digested CO387 and CG62 genomic DNA samples were probed with the DIG labeled 420 bp PIC11 fragment (Collins *et al.*, 1998) that gave a single band of different sizes. Parent CO387 sample had upper band whilst parent CG562 samples had lower band (Fig. 2A). The RILs derived from a cross between these inbred lines had either the parental patterns or heterozygote for the two patterns (Fig. 2A).

A SCAR marker was developed for the *Zmnbslrr1* sequence that showed a 2100 bp band in CO387 and no band in CG62 (Fig. 2B).

An analysis of the predicted amino acid sequence for the ZmNBSLRR1 protein indicated that it shared a large number of homologies with NBS-LRR-like disease resistance genes found in rice, barley, wheat, and other crops. Alignment of conserved regions among resistance gene products revealed that the putative amino acid residues of ZmNBSLRR1 matched the consensus sequence found among resistance genes. Fig. 3A showed the significant similarity between the ZmNBSLRR1 amino acid sequence and the products of NBS-LRRs resistance genes such as RPR1 (BAA75812; Sakamoto *et al.*, 1999), LRR19 (AAK20736; Feuillet *et al.*, 2001), RXO1 (AAX31149; Zhao *et al.*, 2005), OS-NB-ARC (ABA92222; Buell *et al.*, 2005) and HV-NBS-LRR (CAD45036; Madsen *et al.*, 2003). The ZmNBSLRR1 possessed an ATP/GTP-binding motif (Traut, 1994) in the N-terminus that was composed of a P-loop domain (Saraste *et al.*, 1990) at residues 201–211, a kinase 2 domain at 278–290 and followed by PLGL domain at 371–383. However, the *Zmnbslrr1* showed a reduced domain in the C-terminus and imperfect LRRs (Fig. 3B). Those similar genes all belonged to the CC-NBS-LRR gene family. Clustal analysis demonstrated that *Zmnbslrr1* was not most closely related with RPR1. Since RPR1 was induced by probenazole (Fig. 4), *Zmnbslrr1* may be more functionally close to barley NBS-LRR (Madsen *et al.*, 2003).

Our previous study (Yuan *et al.*, 2008) showed that genomic DNA of *Zmgc1* gene spanned 3.3 kb while cDNA clones were 1.3 kb and contained a predicted protein of 267 amino acids (Yuan *et al.*, 2008). In the upstream of the gene coding sequence, the gene contained GC-rich sequences and no obvious TATA box was identified. The complete nucleotide sequence consisted of nine exons

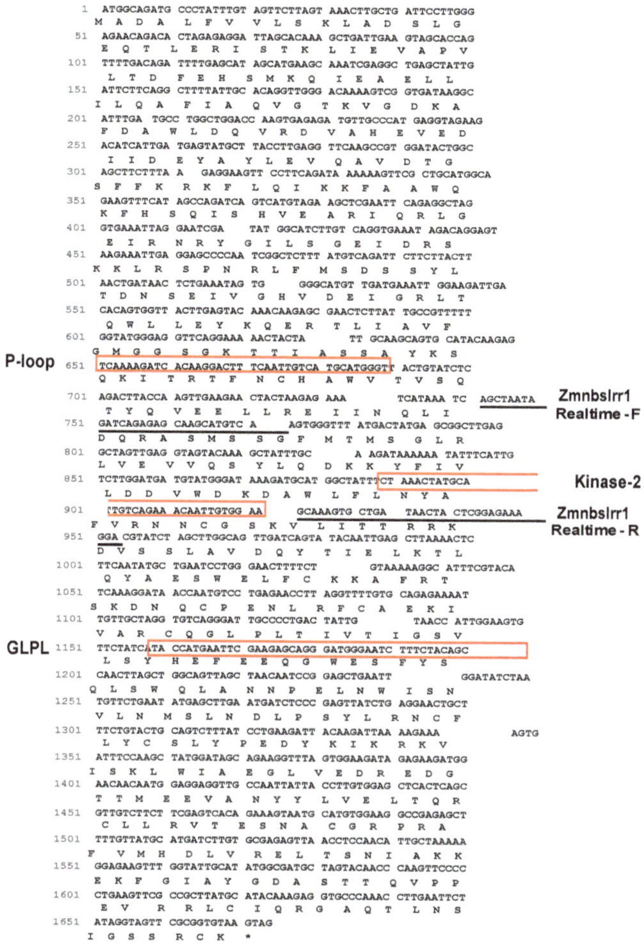

Fig. 1. ORF of *Zmnbslrr1* and the deduced amino acid sequences from genotype CO387. The amino acid coding sequence was shown by the one-letter code. The positions of the primers (Zmnbslrr1-Realtime-F and Zmnbslrr1-Realtime-R) for real-time PCR were marked by black arrows. The alignment was performed using ClustalX (Thompson *et al.*, 1997).

interrupted by eight introns evenly distributed within the gene. The ortholog in *Arabidopsis* AtGC1 (GenBank accession number AAM51559) was the most similar GC known to ZmGC1and 53% identical and 65% similar to ZmGC1. Deduced amino acid sequence of ZmGC1 was aligned with those of GCs that had been used to search common motifs (Lukiki and Gehring, 2003; Yuan *et al.*, 2008). However, *Zmnbslrr1* was not assigned any *Fusarium* resistance in our mapping population.

Genetic mapping and disease resistance of the Zmnbslrr1 and Zmgc1

PIC11 marker was positioned on the linkage group 5 of developed previously for genetic map of the population and 9 cM and 16 cM from molecular markers umc1056 and BC399_1300 respectively based on RFLP data (Fig. 5). The SCAR (sequence characterized amplified region) marker (Fig. 2B) derived from total GenomeWalking sequence containing *Zmnbslrr1* gene sequence also

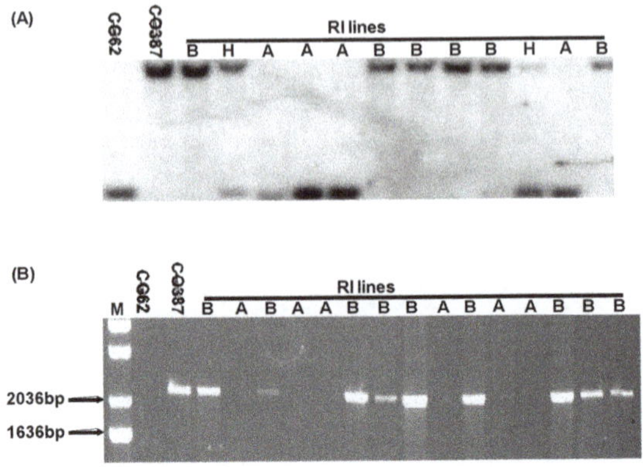

Fig. 2. Two molecular mapping approaches to position the *Zmnbslrr1* gene on the linkage map derived from F_5 RIs crossed from CG62 x CO387. **A** X-ray films exposed to membranes of the F5 population genomic DNA hybridized with a probe derived from sequence information of *Zmnbslrr1* gene. The probe was amplified using primers RFLP-PIC11-F 5′-CTAATAGATCAGAGAGCAAGCA-3′ and RFLP-PIC11-R 5′-GTTCCCAGGATTCAGCATATTG-3′ labeled with DIG method (Boehringer Mannheim, Germany). Recombinant inbred (RI) lines were digested with restriction enzyme *Eco*RV and one RFLP co-dominant marker was scored. A: genotype same as CG62, B: genotype same as CO387, H: heterozygous. **B** Sequence-characterized amplified region (SCAR) marker of PIC11. The fragments were amplified from parental DNA CO387 and CG62 and RI lines using primers (primer SCAR-PIC11-F 5′-ACTGGGCATGAAGGGTACTAG-3′ and SCAR-PIC11-R 5′- CTTAATCAGGTCCCTGGTTAGT-3′). The map position of the SCAR marker is the same as that of RFLP. A: genotype same as CG62, B: genotype same as CO387.

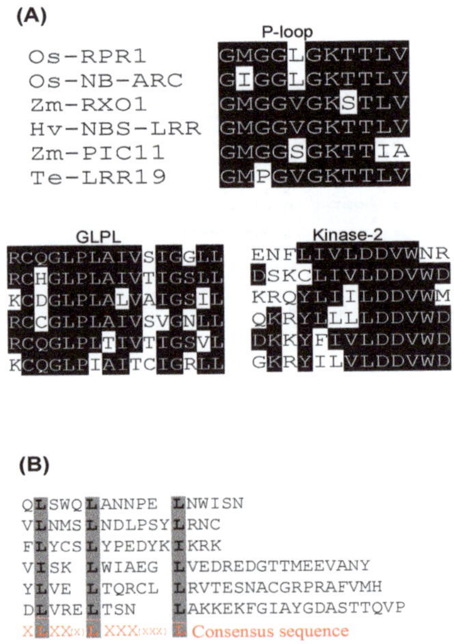

Fig. 3. Alignment of the deduced amino acid of ZmNBSLRR1 with other NBS-LRR genes. **A** Similarity between the ZmNBSLRR1 amino acid sequence and the products of NBS-LRRs resistance genes such as the rice RPR1 (BAA75812; Sakamoto *et al.*, 1999), LRR19 (AAK20736; Feuillet *et al.*, 2001), RXO1 (AAX31149; Zhao *et al.*, 2005), OS-NB-ARC (ABA92222; Buell *et al.*, 2005) and HV-NBS-LRR (CAD45036; Madsen *et al.*, 2003). The ZmNBSLRR1 possesses an ATP/GTP-binding motif in the N-terminus that was composed of a P-loop domain, a kinase 2 domain, and followed by PLGL domain. **B** ZmNBSLRR1 showed obviously a reduced size domain in the C-terminus and it contained imperfect LRRs.

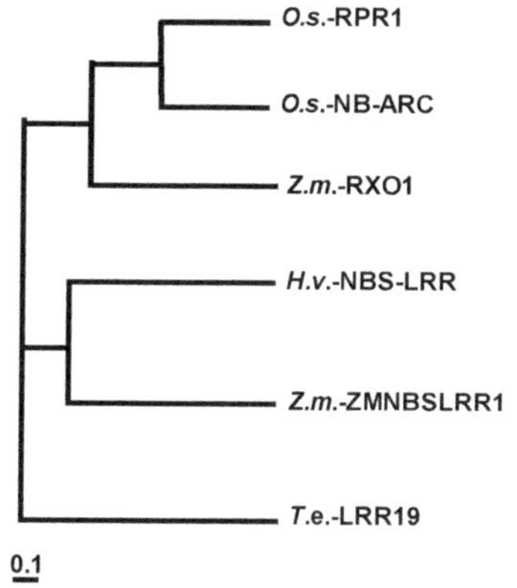

Fig. 4. Phylogenetic analysis of ZmNBSLRR1. Amino acid sequences of NBS-LRRs and putative amino acid sequence of ZmNBSLRR1 were aligned with the Clustal X program (Thompson *et al.*, 1997). The scale bar represents an evolutionary distance of 0.1 amino acid substitution per residue.

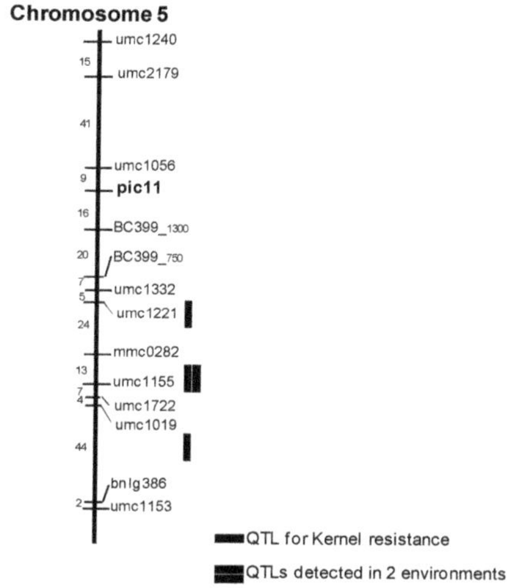

Fig. 5. Map position of the PIC11 on corn chromosomes. The genetic map was generated from segregation data of the CG62 x CO387 recombinant inbred lines using MAPMAKER/EXP 3.01b and QTL Cartographer 2.0. Relative positions of the markers were indicated in cM on the left of chromosomes and the markers were on the right of the chromosome. The extent of the QTL above a LOD threshold of 2.0 is indicated by the length of the bar. QTL detected in more than one environment are indicated by multiple bars.

confirmed the gene position on the genetic map. The mapping position matches that of others (Collins *et al.*, 1998).

Six polymorphic markers derived from *Zmgc1* gene were identified in the membranes containing the individuals digested with restriction enzyme *Eco*RI (ER1_1 to ER1_5) and *Eco*RV (ER5_1). Two RFLP bands, ER1_1 and ER5_1, were significantly associated with gibberella ear rot resistance in maize across different environments in a recombinant inbred (RI) population (Yuan *et al.*, 2008).

Zmnbslrr1 and Zmgc1 gene expression

In order to examine the function of *Zmnbslrr1* and *Zmgc1* gene in maize, the response of the transcript abundances to probenazole was measured. The pattern of mRNA expression of the *Zmgc1* gene showed clear differences compared to that of the *Zmnbslrr1* gene in corn seedling tissue after the probenazole treatment (Fig. 6).

Fig. 6 showed that the basal expression level of *Zmgc1* gene was low in untreated plants and the transcript abundance rapidly reached a peak within 2h after pathogen challenge. In contrast, the reduction of *Zmnbslrr1* gene expression was observed at this early stage. Significantly increased *Zmgc1* transcript abundance (more than 10 folds) was detected 8 hours after treatment with probenazole. The transcription abundance remained higher throughout the time-course study. For six days after the probenazole applied, the abundance of the *Zmgc1* mRNA was still notably higher than that of basal level. However, transcript abundance of *Zmnbslrr1* in the seedling treated with probenazole was reduced substantially by PBZ treatment throughout the time course. The transcript abundance of *Zmnbslrr1* gene was notably lower than that of *Zmgc1* gene.

Discussion

RGAs can be cloned by PCR-based approaches and/or by using genomic information and then genetically mapped. Kojima *et al.*, (2002) demonstrated a successful example using known gene sequences to isolate related genes, in which the QTL had been confirmed as candidate genes whose function was known from the researches of Arabidopsis. Collins *et al.*, (1998) presented one of the molecular markers named PIC11 derived from a RGA. PIC11 RGA had been mapped to a locus that was a previously defined resistance locus. The locus was associated with Gibberella stalk rot that was caused by *Gibberella zeae* (petch), which is the sexual form of *F. graminearum* (Pe *et al.*, 1993). If functionality of the *Zmnbslrr1* gene can be tested, then it will be easy to determine whether the candidate gene confers resistance when it is expressed as a transgene in a susceptible line. The putative amino acid sequence of ZmNBSLRR1 showed the similarity with NBS-LRR genes including RPR1 (BAA75812; Sakamoto *et al.*, 1999), LRR19 (AAK20736; Feuillet *et al.*, 2001), RXO1 (AAX31149; Zhao *et al.*, 2005), OS-NB-ARC (ABA92222; Buell *et al.*, 2005) and HV-NBS-LRR (CAD45036; Madsen *et al.*, 2003). Sakamoto *et al.*, (1999) demonstrated that the gene product, RPR1, contained motifs of a nucleotide binding site (NBS) and a leucine-rich repeats (LRR) domain and RPR1 gene expression was up-regulated by treated with probenazole, chemical inducers of systemic acquired

A

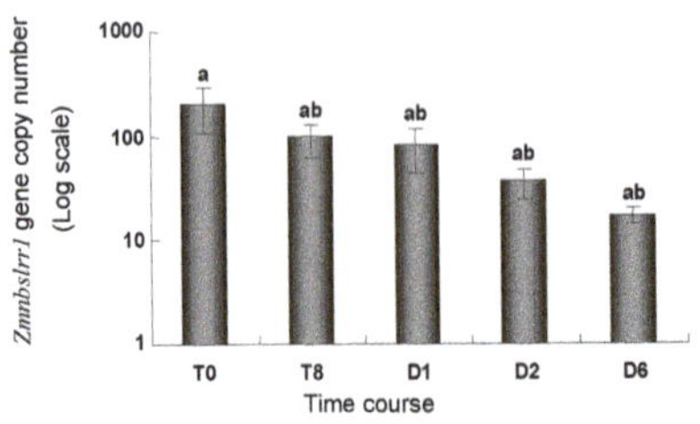

B

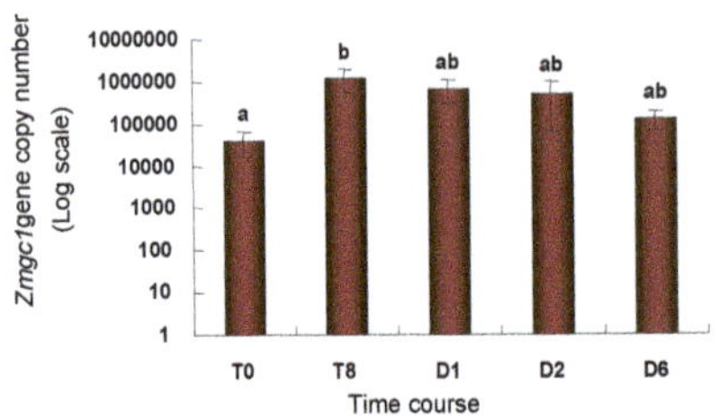

Fig. 6. Real-time PCR analysis for the *Zmnbslrr1* and *Zmgc1* gene transcript abundances in the leave tissue after probenazole treatment. The copy number of the transcript abundances was indicated at the Y axis and the time after probenazole treated was at the X axis. T: hour after the treatment; D: day after the treatment. Dark grey legends indicated the mean transcript abundance of *Zmnbslrr1*-CO387, whereas red legend showed the mean transcript abundance of *Zmgc1*-CO387. **A** The realtime PCR was amplified by primers (Zmnbslrr1-realtime-F 5′-AGCTAATAGATCAGAGAGCAAGCATGTCA-3′ and Zmnbslrr1-realtime-R 5′-TCCTTTCTCCGAGTAGTTATCAGCACTTTG-3′) was about 180 bp. **B** The realtime PCR was amplified by primers (Zmgc1-Realtime-F 5′-ACAACAGCACAACGATGAGCCTGACTAC-3′ and Zmgc1-Realtime-R 5′-CAATGAAGAGCTTGGACGAGGCCCGCAAAT-3′) was approximately 150 bp. Means labeled with the same letter were not significantly different ($P<0.05$).

resistance (SAR) such as BTH and SA, and pathogens. However, PIC11 was mapped to same locus as that of other's study but it was not assigned any resistance in our mapping population. Hulbert *et al.*, (2001) suggested that resistance loci are often comprised of complex multicopy clusters of genes. The duplication events may play an important role in generating new resistance gene specificities in R-gene and RGA clusters. Duplicated RGA sequences result from transposition, unequal crossing-over, gene conversion, and point mutation in several crops (Meyers *et al.*, 1998; Collins *et al.*, 1999; Ellis *et al.*, 1999). However, the reason for the reduced size of *Zmnbslrr1* gene is not known and only a single copy of *Zmnbslrr1* was identified. *Zmnbslrr1* might be a pseudogene due to the size of the gene but the detected transcript suggested that it was expressed.

Probenazole induces systemic acquired resistance (SAR)-like disease resistance in rice (*Oryza sativa*) has been widely used against rice blast disease for many years in Asia (Sakamoto *et al.*, 1999). For plant sustainable disease management, chemical activated disease resistance may be an option for growers to protect their crops from plant diseases if the genetic resources could not provide sufficient resistance. It is necessary to understand about the various defense signal pathways in plants previous application of commercial chemical activators. The differential expression of *Zmnbslrr1* and *Zmgc1* gene before and after probenazole treated suggested that these genes might act different metabolic pathways corresponding to same signal. Induction of *Zmgc1* transcript was the response expected for a gene involved in partial resistance. Reduction in *Zmnbslrr1* transcript abundance may suggest that it is not involved in resistance to *F. graminearum*, but represents a competing pathway or contributes to susceptibility.

In plants, cGMP is involved in modulating gene expression of many defense and stress related pathways either functioning alone or combining with calcium (Bowler *et al.*, 1994). Whether the guanylyl cyclase plays direct roles in perception of host stimuli or has roles in signal processing after probenazole remains to be determined. Gibberella ear rot infections usually occur when the cool weather precedes wet and warm condition during flowering. If the abiotic stress (cold) affects the abundance of cGMP and transcription of the GC gene(s) remains to be addressed. Enhancing activity of cell membrane GTPase, an enzyme involved in intracellular signal transduction pathway had been identified after the rice plant treated with probenazole (Kanoh *et al.*, 1993). Although the relationship of probenazole induced resistance and SAR-like resistance induced by pathogen is still fragmentary, it suggested that the protective effects of the chemical mainly due to activation of host defense mechanism (Sakamoto *et al.*, 1999). If there is any correlation between GTP metabolism and disease resistance or probenazole induction is still unknown. How the *Zmgc1* gene complement interacts with other genes is also not clear, but understanding these processes may lead to novel strategies to interfere in recognition host-pathogen.

Acknowledgments

This work was supported by the Ontario Corn Producers, the Ontario Corn Consortium (including the Ontario Pork Producers, Pioneer Hi-Bred, Syngenta Seeds Canada, Inc., Agrevo, King Agro, Pickseed Canada, W.G. Thompson, and Zeneca), the Ontario Ministry for Food and Agriculture, Ontario Research Enhancement Program, Canadapt, Federation des Producteurs de Cultures Commerciales du Quebec, the Natural Sciences and Engineering Research Council of Canada and Agriculture and Agri-Food Canada.

Angela Hill is gratefully acknowledged for assistance in preparing the manuscript.

References

Ali, L., Taylor, J., Liu, J., Sun, G., William, M., Kasha, K., Reid, L., and Pauls, P. (2005) Molecular mapping of QTLs for resistance to Gibberella ear rot in maize caused by *Fusarium graminearum*. Genome *48*, 521–533.

Belkhadir, Y., Subramaniam, R., and Dangl, L.J. (2004) Plant disease resistance protein signaling: NBS-LRR proteins and their partners. Curr. Opin. in Plant Biol. *7*, 391–199.

Berridge, M.J. (1993) Cell signaling. A tale of two messengers. Nature *365*, 388–389.

Bowler, C., Neuhaus, G., Yamagata, H., and Chua, N.H. (1994) Cyclic GMP and calcium mediate phytochrome phototransduction. Cell *77*, 73–81.

Buell, C.R., Wing, R.A., McCombie, W.A., and Ouyang, S. (2005) The sequence of rice chromosomes 11 and 12, rich in disease resistance genes and recent gene duplications. BMC Biol *3*, 20–20.

Christensen, C.M., and Meronuck. R.A. (1986) Quality maintenance in stored grains and seeds. (Minneapolis, MN: University of Minneapolis Press).

Chungu, C., Mather, D.E., Reid, L.M., and Hamilton, R.I. (1996) Inheritance of kernel resistance to *Fusarium graminearum* in maize. J. Heredity *87*, 382–385.

Clay, K., and Kover, P.X. (1996) The Red Queen hypothesis and plant:pathogen interactions. Annu. Rev. Phytopathol. *34*, 29–50.

Collins, N., Webb, C., Seah, S., Ellis, J., Hulbert, S., and Pryor, A. (1998) The isolation and mapping of disease resistance gene analogs in maize. Mol. Plant Microbe Interact. *11*, 968–978.

Covert, S.F., Kapoor, P., Lee, M., Briley, A., and Nairn, C.J. (2001) Agrobacterium tumefaciens-mediated transformation of Fusarium circinatum. Mycol. Res. *105*, 259–264.

Dangl, J.L., and Jones, J.D. (2001) Plant pathogens and integrated defence responses to infection. Nature *411*, 826–833.

Davis, G., McMullen, M., Baysdorfer, C., Musket, T., Grant, D., Staebell, M.S., Xu, G., Polacco, M., Koster. L., Melia-Hancock, S., Houchins, K., Chao, S., and Coe, H.E. (1999) A maize map standard with sequenced core markers, grass genome reference points, and 932 ESTs in a 1736- locus map. Genetics *152*, 1137–1172.

Desjardins, A.E. (2003) Trichothecenes: from yellow rain to green wheat. ASM News *69*, 182–185.

Durner, J., Wendehenne, D., and Klessig, D. (1998) Defense gene induction in tobacco by nitric oxide, cyclic GMP, and cyclic ADP ribose. Proc. Natl. Acad. Sci. U.S.A. *95*, 10328–10333.

Ellis, J.G., Lawrence, G.J., Luck, J.E., and Dodds, P.N. (1999) The identification of regions in alleles of the flax rust resistance gene L that determine differences in gene-for-gene specificity. Plant Cell *11*, 495–506.

Feuillet, C., Penger, A., Gellner, K., Mast, A., and Keller, B. (2001) Molecular evolution of receptor-like kinase genes in hexaploid wheat. Independent evolution of orthologs after polyploidization and mechanisms of local rearrangements at paralogous loci. Plant Physiol. *125*, 1304–1313.

Grant, M.R., Godiard, L., Straube, E., Ashfield, T., Lewald, J., Sattler, A., Innes, R.W., and Dangl, J.L. (1995) Structure of the Arabidopsis RPM1 gene enabling dual specificity disease resistance. Science *269*, 843–846.

Greenhalgh, R., Neish, G.A., and Miller, J.D. (1983) Deoxynivalenol, acetyl deoxynivalenol, and zearalenone formation by Canadian isolate of *Fusarium graminearum* on solid substrates. Appl. Environ. Microbiol. *46*, 625–629.

Hoisingtonm, D.A., Khairallah, M.M., and Gonzales-de-Leon, D. (1994) Laboratory protocols. CIMMYT Applied Molecular Genetics Laboratory. (Hisfoa, Mexico: CIMMYT)

Hulbert, S.H., Webb, C.A., Smith, S.M., and Sun, Q. (2001) Resistance gene complexes: Evolution and utilization. Annu. Rev. Phytopathol. *39*, 285–312.

Iqbal, M.J., Yaegashi, S., Ahsan, R., Shopinski, K.L., and Lightfoot, D.A. (2005) Root response to *Fusarium solani* f. sp . glycines: temporal accumulation of transcripts in partially resistant and susceptible soybean. Theor. Appl. Genet. *110*, 1429–1438.

Kanoh, H., Haga, M., Iwata, M., and Sekizawa, Y. (1993) Transmembrane signaling operated at rice blade cells stimulated by blast fungus elicitor I: Operation of the phosopholipase C system. Journal of Pesticide Science *18*, 299–308.

Kojima, S., Takahashi, Y., Kobayashi, Y., Monna, L., Sasaki, T., Araki, T., and Yano, M. (2002) Hd3a, a rice ortholog of the Arabidopsis FT gene, promotes transition to flowering downstream of Hd1 under short-day condition. Plant Cell Physiol. *43*, 1096–1105.

Kosambi, D.D. (1944) The estimation of map distance from recombination values. Ann. Eugen. *12*, 172–175.

Lander, E.S., Green, P., Abrahamson, J., Barlow, A., Daly, M.J., Lincoln, S.E., and Newburg, L. (1987) MAPMAKER: an interactive computer package for constructing primary genetic linkage maps of experimental and natural populations. Genomics *1*, 174–181.

Langcake, P., Kuhn, P.J., and Wade, M. (1983) The mode of action of systemic fungicides. In: Hutson, D.H., Roberts, T.R. ed. (Progress in pesticide biochemistry and toxicology. Chichester, John Wiley) pp. 1–109.

Levine, A., Pennell, R.I., Alvarez, M.E., Palmer, R., and Lamb, C. (1996) Calcium-mediated apoptosis in a plant hypersensitive disease resistance response. Curr. Biol. *6*, 427–437.

Lee, E.A., Good, B., Chakravarty, R., and Kannenberg, L. (2001) Corn inbred lines CG60 and CG62. Can. J. Plant Sci. *81*: 453–454.

Ludidi, N., and Gehring, C. (2003) Identification of a novel protein with guanylyl cyclase activity in *Arabidopsis thaliana*. J. Biol. Chem. *278*, 6490–6494.

Madsen, L.H., Collins, N.C., Rakwalska, M., Backes, G., Sandal, N., Krusell, L., Jensen, J., Waterman, E.H., Jahoor, A., Ayliffe, M., Pryor, A.J., Langridge, P., Schulze-Lefert, P., and Stougaard, J. (2003) Barley disease resistance gene analogs of the NBSLRR class: identification and mapping. Mol. Gen. Genomics *269*, 150–161.

Meyers, B.C., Shen, K.A., Rohani, P., Gaut, B.S., and Michelmore, R.W. (1998) Receptor-like genes in the major resistance locus of lettuce are subject to divergent selection. Plant Cell *11*, 1833–1846.

Meyers, B.C., Kozik, K., Griego, A., Kuang, H., and Michelmore, R.W. (2003) Genome–wide analysis of NBS-LRR encoding genes in Aarbidopsis. Plant Cell *4*, 809–834.

Michelmore, R. (2000) Genomic approaches to plant disease resistance. Curr. Opin. Plant Biol. *3*, 125–131.

Molto, G.A., Gonzalez, H. H.L., Resnik, S.L., and Gonzalez, A.P. (1997) Production of trichothecenes and zearalenone by isolates of Fusarium ssp. from Argentinean maize. Food Addit. Contam. *14*, 263–268.
Oostendorp, M., Kunz, W., Dietrich, B., and Staub, T. (2001) Induced resistance in plants by chemicals. Euro. J. Plant Pathol. *107*, 19–28.
Pe, M.E., Gianfranceschi, L., Taramino, G., Tarchini, R., Angelini, P., Dani, M., and Binelli, G. (1993) Mapping quantitative trait loci (QTLs) for resistance to *Gibberella zeae* infection in maize. Mol. Gen. Genet. *241*, 11–16.
Pink, D.A.C. (2002) Strategies using genes for non-durable resistance. Euphytica *1*, 227–236.
Reid, L.M., Mather, D.E., Bolton, A.T., and Hamilton, R.I. (1994) Evidence for a gene for silk resistance to *Fusarium graminearum* Schw. ear rot of maize. J. Hered. *85*, 118–121.
Reid, L.M., and Hamilton, R.I. (1996) Effect of inoculation position, timing, macroconidial concentration, and irrigation of resistance of corn to *Fusarium graminaerum* infection through kernels. Can. J. Plant Pathol. *18*, 279 – 285.
Rossi, M., Goggin, F.L., Milligan, S.B., Kaloshian, I., Ullman, D.E., and Williamson, V. M. (1998) The nematode resistance gene Mi of tomato confers resistance against the potato aphid. Proc. Natl. Acad. Sci. U.S.A. *95*, 9750–9754.
Ryals, J.A., Neuenschwander, U.H., Willits, M. G, Molina, A., Steiner, H., and Hunt, D.M. (1996) Systemic acquired resistance. Plant Cell *8*, 1809–1819.
Sakamoto, K., Tada, Y., Yokozeki, Y., Akagi, H., Hayashi, N., Fujimura, T., and Ichikawa, N. (1999) Chemical induction of disease resistance in rice is correlated with the expression of a gene encoding a nucleotide binding site and a leucine-rich repeat. Plant Mol. Biol. *40*, 847–855.
Smale, S.T. (1997) Transcription initiation from TATA-less promoters within eukaryotic protein-coding genes. Biochem. Biophys. Acta *1351*, 73–88.
Saraste, M., Sibbald, P,R,, and Wittinghofer, A. (1990) The P-loop – a common motif in ATP- and GTP-binding proteins. Trends Biochem. Sci. *15*: 430–4.
Sutton, J.C. (1982) Epidemiology of wheat head blight and maize ear rot caused by *Fusarium graminearum*. Can. J. Plant Pathol. *4*, 195 – 209.
Thompson, J.D., Higgins, D.G., and Gibson, T.J. (1994) CLUSTAL W: improving the sensitivity of progressive multiple sequence alignment through sequence weighting, position-specific gap penalties and weight matrix choice. Nucleic Acids Res. *22*, 4673–4680.
Thompson, J.D., Gibson, T.J., Plewiak, F., Jeanmougin, F., and Higgins, D.G. (1997) The CLUSTAL_X windows interface: flexible strategies for multiple sequence alignment aided by quality analysis tools. Nucleic Acids Res. *24*, 4876–4882.
Traut, T.W. (1994) The functions and consensus motifs of nine types of peptide segments that form different types of nucleotide-binding sites. Eur. J. Biochem. *222*, 9–19.
Uchiyama, M., Abe, H., Sato, R., Shimura, M., and Watanabe, T. (1973) Fate of 3-allyloxy-1,2-benzisothiazole 1,1-dioxide (Oryzemater) in rice plants. Agricultural and Biological Chemistry. *37*, 737–745.
Wang, S., Basten, C.J., and Zeng, Z.B. (2006) Windows QTL Cartographer 2.5. (Raleigh, NC.)
Watanabe, T., Sekizawa, Y., Shimura, M., Suzuki, Y., Matsumoto, K., Iwata, M., and Mase, S. (1979) Effects of probenazole (Oryzemate®) on rice plants with reference to controlling rice blast. Journal of Pesticide Science *4*, 53–59.
Whitham, S., Dinesh-Kumar, S.P., Choi, D., Hehl, R., Corr, C., and Baker, B. (1994) The product of the tobacco mosaic virus resistance gene N: Similarity to toll and the interleukin-1 receptor. Cell *78*, 1101–1115.
Yuan, J., Ali, L., Taylor, J., Liu, J., Sun, S., Liu, W., Masilimany, P., Gulati-Sakhuja, A., and Pauls, K.P. (2008) A guanylyl cyclase-like gene is associated with Gibberella ear rot resistance in maize (Zea mays L.) Theor. Appl. Genet. *116*, 465–479.
Zhao, B., Lin, X., Poland, J., Trick, H., Leach, J., and Hulbert. S. (2005) A maize resistance gene functions against bacterial streak disease in rice. Proc. Natl. Acad. Sci. U.S.A. *102*, 15383–15388.